EMPOWERING CIVIL SOCIETY IN THE INDUSTRIAL REVOLUTION 4.0

PROCEEDINGS OF THE 1ST INTERNATIONAL CONFERENCE ON CITIZENSHIP EDUCATION AND DEMOCRATIC ISSUES (ICCEDI 2020), MALANG, INDONESIA, OCTOBER 14, 2020

EMPOWERING CIVIL SOCIETY IN THE INDUSTRIAL REVOLUTION 4.0

Edited by

Sukowiyono
Universitas Negeri Malang, Indonesia

Siti Awaliyah
Universitas Negeri Malang, Indonesia

Peter B.R. Carey
University of Oxford, UK

Max Lane
The Institute of Southeast Asian Studies, Australia

Chen Fei
School of Politics and International Studies, Normal University, Central China

Routledge
Taylor & Francis Group

Routledge is an imprint of the Taylor & Francis Group, an informa business

Typeset by MPS Limited, Chennai, India

Library of Congress Cataloging-in-Publication Data

A catalog record has been requested for this book

First published 2021
by Routledge
2 Park Square, Milton Park, Abingdon, Oxon OX14 4RN
and by Routledge
605 Third Avenue, New York, NY 10158
e-mail: enquiries@taylorandfrancis.com
www.routledge.com – www.taylorandfrancis.com

ISBN: 978-1-032-01801-0 (Hbk)
ISBN: 978-1-032-01805-8 (Pbk)
ISBN: 978-1-003-18012-8 (eBook)
DOI: 10.1201/9781003180128

Empowering Civil Society in the Industrial Revolution 4.0 – Sukowiyono et al (Eds)

Table of contents

Empowering Civil Society in the Industrial Revolution 4.0 – Sukowiyono et al (Eds)
© 2021 copyright the Author(s), ISBN 978-1-032-01801-0.
Open Access: www.taylorfrancis.com, CC BY-NC-ND 4.0 license

Preface

The International Conference on Citizenship Education and Democratic Issues (ICCEDI) 2020 was held at the Universitas Negeri Malang in East Java, Indonesia on October 14, 2020. This Conference was brought by the Law and Citizenship Departement, Faculty of Social Science of the Universitas Negeri Malang. There were four main speakers in this conference: Prof. Murray Print, PhD from the University of Sidney, Australia, Prof. Jeffrey A. Winters from Northwestern University, United States, Hery Yanto The, PhD from Zhejiang Yuexiu University, China, and Dr. Didik Sukriono, S.H., M.H from Universitas Negeri Malang, Indonesia.

The committee received 116 papers of which 56 papers were accepted to be presented. Participants from various universities in Indonesia and some other countries made the conference truly international in scope. Of the total number of presented papers, 34 papers were accepted to be included in the book series published by CRC Press Balkema, Taylor and Francis Group.

Generous support for the conference was provided by the Law and Citizenship Departement, Faculty of Social Science, Universitas Negeri Malang, and the Rector of Universitas Negeri Malang for financial support of the event. The ICCEDI 2020 was successfully held and we invite presenters around the world to participate in the next ICCEDI which will be held in 2022. Finally, given the rapidity of civil society, we hope that the future ICCEDI will be as stimulating as indicated in this proceedings volume.

Empowering Civil Society in the Industrial Revolution 4.0 – Sukowiyono et al (Eds)

Acknowledgment

All of the papers in this book are the final version of the manuscripts that were presented at the International Conference on Citizenship Education and Democratic Issues (ICCEDI), held on October 14, 2020 in Malang, Indonesia. We would like to thank the Rector, Dean of Faculty of Social Science and Head of Law and Citizenship Departement of Universitas Negeri Malang, and all of the staff for their great support. On behalf of the organizers, authors, and readers, we also wish to acknowledge all keynote speakers and reviewers for their hard work, time, and dedication to this conference. Their effort maintains the high standard of all the papers in this book. Our thanks also go to the participants and presenters of the conference. Finally, many thanks are given to all the committee members and everyone that helped and supported this conference.

The organizers wish to apologize to the presenters who cannot publish their paper in this conference proceeding. Our apology also goes to all the participants for all the shortcomings in this conference. See you at the next 2nd ICCEDI at Universitas Negeri Malang.

Malang, March 25, 2021

Organizer of ICCEDI
Universitas Negeri Malang

Empowering Civil Society in the Industrial Revolution 4.0 – Sukowiyono et al (Eds)

Organizing committee ICCEDI 2020

Advisory Board
Prof. Dr. AH. Rofi'uddin, M.Pd
Universitas Negeri Malang, Indonesia

Chairman
Muhammad Mujtaba Habibi, S.Pd., M.AP
Universitas Negeri Malang, Indonesia

Secretary
Sudirman, S.Pd., M.H
Rista Ayu Mawarti, S.Pd., M.Pd
Universitas Negeri Malang, Indonesia

Treasurer
Dr. Siti Awaliyah, S.Pd., M.Hum
Arif Wicaksono, S.T., M.M
Universitas Negeri Malang, Indonesia

Technical Chair
Mifdal Zusron Alfaqi, S.Pd., M.Sc
Abd. Mu'id Aris Shofa, S.Pd., M.Sc
Desinta Dwi Rapita, S.Pd., M.H
Rani Prita Prabawangi, S. Hub. Int, M.Si
Neo Adhi Kurniawan, S.Pd., S.H., M.H
Surya Desismansyah Eka Putra, S.Pd., M.Phil
Hanafi, S.Pd., M.Pd
Mulyawati
Universitas Negeri Malang, Indonesia

Publication
Meidi Saputra, S.Pd., M.AP
Idris, S.S., M.M.

Empowering Civil Society in the Industrial Revolution 4.0 – Sukowiyono et al (Eds)
© 2021 copyright the Author(s), ISBN 978-1-032-01801-0.
Open Access: www.taylorfrancis.com, CC BY-NC-ND 4.0 license

Scientific committee ICCEDI 2020

Saida Ulfa, M.Edu, PhD
Universitas Negeri Malang, Indonesia

Aji Prasetya Wibawa, PhD
Universitas Negeri Malang, Indonesia

Prof. Dr.Sapriya, M. Pd
Universitas Pendidikan Indonesia, Indonesia

Dr. Arqom Kuswanjono
Universitas Gadjah Mada, Indonesia

Prof. Peter Carey
University of Oxford, UK

Dr. Max Lane
The Institute of Southeast Asian Studies, Australia

Assoc. Prof. Chen Fei, PhD
Central Normal China University, China

Empowering Civil Society in the Industrial Revolution 4.0 – Sukowiyono et al (Eds)

Efforts to increase participation of beginner voters in the general election of 2019

D. Sukriono*
Universitas Negeri Malang, Malang, Indonesia

ABSTRACT: This article describes the efforts made by the government to increase the participation of beginner voters in the simultaneous general election held in 2019. The study uses a qualitative approach with a descriptive type. Data was collected through interviews, observations, and documentation. Data was analyzed using an interactive model with three activities flowing together, namely reducing data, presenting data and drawing conclusions. The problems faced by beginner voters as one of the determinants of the success of the election are very diverse, one of which is the lack of information regarding the implementation of elections and the problem of administrative requirements that must be met by beginner voters. Efforts made by the General Election Commission (*KPU*) to increase the participation of beginner voters include socialization, optimization of General Election Commission social media, the formation of democratic volunteers, and providing an election smart home. This is also a major factor that can affect the level of voter turnout in simultaneous elections.

Keywords: beginner voters, participation, simultaneous elections, socialization

1 INTRODUCTION

The election implies the implementation of governance mechanisms in an orderly, orderly and peaceful manner and the birth of a society that can respect the opinions of others. Besides that, a society that has a high level of criticism will be born, in the sense that it is selective or usually chooses the best according to its beliefs (KPU, 2015).

However, in reality there are still people who are ignorant of the electoral process and do not even participate in elections that are held (white group/abstentions). Based on the Central Election Commission (KPU) notes, the abstention rate from year to year is increasing, although in 2014 there was a slight increase. The data shows that abstentions in consecutive elections since 1999, 2004, 2009 and 2014 reached 6.70%, 15.93%, 29.01% and 24.89% (Irawan, 2014).

Elections are held to elect members of the Parliament, Regional Representative Council, President and Vice President and Regional Parliament. This is in accordance with Article 22E paragraph (2) of the 1945 Constitution of the Republic of Indonesia: "General elections are held to elect members of the parliament, Regional Representative Council, the President and Vice-President, and the Regional Parliament".

Decision of the Constitutional Court Number 14/PUU-XI/2013 related to the implementation of the election of President and Vice President (hereinafter referred to as the Presidential Election) and the Election of Members of the Parliament, Regional Representative Council, Regional Parliament (hereinafter referred to as legislative election), which have been carried out simultaneously, have brought a new nuance in the democratic process in Indonesia (Rapita, 2014).

*Corresponding Author

DOI 10.1201/9781003180128-1

Law No. 7 of 2017 concerning general elections provides guarantees for beginner voters who, on 17 April 2019, are at least 17 years old to be able to exercise their voting rights in the 2019 elections. Of course, if we count the number, it is large compared to the previous elections. Around 196.5 million voters in the 2019 elections, and 7.4 percent of them or around 14 million voters are young people who have the right to vote for the first time (Soesatyo, 2019). Whereas the Ministry of Home Affairs recorded about 5 million beginner turnouts in the 2019 elections. This data was obtained from the Potential Electoral Population Register (*DP4*). Delivered by the Director General of Population and Civil Registration of the Ministry of Home Affairs, Fakrulloh (2018), that in the *DP4* there are beginner voters who will be 17 years old from 1 January 2018 to 17 April 2019 with 5,035,887 inhabitants.

2 METHODS

The study uses a qualitative approach with a descriptive type, namely by describing the entirety of the efforts made by the KPU to increase the participation of beginner voters. Specifically, the study was conducted in Malang City. Research locations in the KPU Malang City, Malang City Election Supervisory Board and Malang City community were used as sampling. The reason is that Malang City in 2019 is one of the participants in the regional head elections which is organized according to the simultaneous election agenda. In addition, seeing several previous regional head election periods the community participation rate has decreased. Primary and secondary data was collected through interviews, observations, and documentation (Moleong, 2007). The data obtained was then analyzed using an interactive model with three activities flowing together, namely reducing data, presenting data and drawing conclusions (Miles & Huberman, 2009).

3 RESULTS AND DISCUSSION

3.1 *Problems of beginner voters in 2019 concurrent elections*

Beginner voters according to Wardhani (2018), are voters who have different natures and characters, backgrounds, experience and challenges with voters in the previous generation. Most of them are students, have good economic status, and generally live-in urban areas or surrounding areas. This group is very touched by the advancement of information technology, they use the tools of sophisticated technology well, ranging from cell phones, laptops, tablets and various other gadgets. They are also very fluent in using social media facilities and networks, such as Twitter, Facebook, LinkedIn, and so on.

In general, the problems that occur in the beginner voter in the 2019 simultaneous elections include: *first*, the first voter has not done the recording and printing of the electronic identity card (they do not have electronic identity card). Beginner voters, especially students, still haven't done the recording and printing of electronic identity cards which are used as one of the requirements to become a voter. The number of first-time voters (aged 17 years to 21 years) was 10,310 people, while the number of voter list in Malang City in the 2019 simultaneous elections was 263,185 people (Asminingtyas, 2019); *Second,* the beginner voters can fully use their voting rights even though administratively they can work. The requirements for recording, issuing and granting electronic identity cards can only be done on the day when the population is 17 years old; *Third,* in Law Number 7 of 2017 concerning Elections, an additional Voter List (*DPTb*) is regulated. Or, voters who have registered in the voter list at the polling station who due to certain circumstances the voters cannot use their rights to vote at the polling station where they registered and vote at another polling station. Beginner voters especially do not understand the procedure of moving the region selection correctly; *Fourth,* beginner voters are less interested in election activities. The proof is that many of them do not understand the electoral process and there are even some who do not know and care about the 2019 elections.

3.2 *Participation of beginner voters in 2019 concurrent elections*

Political participation according to Closky inside Budiardjo (2008) is a voluntary activity of community members through which they take part in the process of electing the authorities and directly or indirectly in the general policy formation process. An important aspect of democracy is political participation. Political participation is a hallmark of political modernization.

According to Conyers (1994), there are three main reasons why community participation has very important properties. *First,* community participation is a useful tool to obtain information about the conditions, needs and attitudes of the local community, which are without its presence in development programs and projects will fail. *Second,* the community will trust the development project or program more if it feels involved in the process preparation and planning, because they will know more about the ins and outs of the project and will have a sense of ownership of the project. *Third,* the assumption arises that is a democratic right if the community is involved in community development themselves.

Max Weber in Damsar (2010), the reason people do political activity is because: *first,* for rational reasons and values, namely reasons which are based on rational acceptance of a values group. *Second,* emotional reasons are effective, that is, the reason is based on hateful or voluntary feelings towards ideas, organizations, parties or individuals. *Third,* traditional reasons, namely reasons which are based on acceptance of individuals or norms of behavior in certain traditions of a group social. *Fourth,* rational reasons are instrumental, namely the reasons for which based on the calculation of profit and loss economically.

Based on the data obtained, the level of voter participation in Malang in the 2019 simultaneous elections can be seen in the following table 1.

Table 1. Voter participation rate in Malang city (data obtained from General Election Commission of Malang city)

Types of General Election	Participation Level
President and Vice President	84,08%
Parliament Members	82,21%
Regional Representative Council Member	82,63%
Province Parliament Members	82,21%
Regional Parliament Members	81,99%

This level of community participation has exceeded the target of 77.5%. So that it can be said that the Simultaneous Election in Malang City has been successful and successfully implemented. The high political participation of novice voters in Malang in the 2019 Election, is identical to Milbrath in Sastroatmodjo (1995), namely: (1) acceptance of political incentives, namely openness and acceptance of someone against political incentives through personal contacts, organizations and social media; (2) a person's social characteristics, namely economic status, ethnic characteristics, age, gender and religion. Although novice voters have different characteristics, there are quite a number of novice voters who are concerned and aware of their political rights in society; (3) the political system and party system in which an individual is located. Indonesia as a democracy and there is a tendency of citizens to participate in politics, namely elections; and (4) regional differences, namely environmental aspects that influence differences in the character and behavior of individuals in political participation.

While the obstacles to political participation according to Mas'oed (2008) are: (1) the bustle of daily activities because novice voters are students and workers. Novice voters between the ages of 17 to 21 years are busy working as students and this can make them reluctant to participate in politics; (2) awareness, meaning that the level of education is low or the lack of knowledge in the political field and the low socio-economic life also affects the activeness of a person in the political field; and (3) the family has a great influence on one's political activities. The family can support or even oppose the behavior of family members.

3.3 *Efforts made by KPU to increase participation of beginner voters in 2019 concurrent elections*

James W. Vander Zanden in (Damsar, 2010) defines socialization as the process of social interaction with which people gain knowledge, attitudes, values, and essential behavior for effective participation in society. Furthermore, Dharmawan (2016) defines socialization as the process of planting or transferring habits or values and rules from one generation to another in a group or community.

Normatively general elections are held with public participation as stated in Law Number 7 of 2017 concerning General Elections, namely Chapter XVII Society Participation Article 448: (1) elections are held with public participation; (2) community participation as referred to in paragraph (1) can be done in the form of: a. election socialization; b. political education for voters; c. surveys or polls about elections; and d. quick calculation of election results; (3) the form of community participation referred to in paragraph (2) with the provisions: a. they cannot have alignments beneficial or detrimental to election contestants; b. they cannot interfere with the process of organizing election stages; c. they must aim to increase political participation society at large; and d. they must encourage the realization of a conducive atmosphere for holding a safe, peaceful election orderly and smoothly.

The General Election Commission issued KPU Regulation Number 10 of 2018 concerning Socialization, Voter Education, and Community Participation in the Implementation of General Elections Section 2 Implementation of Socialization, Voter Education, and Participation the community in organizing general elections is guided by the principles of: independence, honesty, legal certainty, order, public interest, openness, proportional, professionality, accountability, efficiency, effectiveness, and accessibility.

The point is according to Dharmawan (2016), the socialization conducted by KPU is considered more focused on the D-day only and lacks an awareness of the importance of the substance the election itself, which is about what, who and how quality and elections have integrity. Therefore, the socialization innovation regarding elections is very much important to do so that the election from time to time gets more quality. After all the election has become a price die as a mechanism to produce democratic government.

Efforts specifically made to increase the participation of novice voters include the Malang City General Election Commission together with the Malang City Election Supervisory Board. In particular the Malang City KPU has carried out various activities including, socialization, optimization of social media General Election Commission, the formation of democratic volunteers and providing Election Smart Houses.

a. Socialization

The General Election Commission cooperates with various parties in socialization activities. In addition, students are the right targets because they will later determine and continue the success of the democratic system in Indonesia, one of which is through their participation in elections. Some universities that have collaborated with the General Election Commission include: Malang State University, Widyagama University of Malang, Brawijaya University, UIN Maulana Malik Ibrahim, Muhammadiyah University of Malang, Malang Islamic University.

Some of the socialization activities carried out by the General Election Commission include: (1) political education and Socialization of the 2019 General Election for Beginner Voters in Schools and held in hotels, (2) making the "*Malang Run Election Mbois*" event in the car free day area, (3) music concert "105 Unity in Harmony", and (4) Seminar on Election Management to students of the Faculty of Social and Political Sciences, Brawijaya University.

b. Optimization of the social media General Election Commission

The General Election Commission of Malang City creates some platforms of social media accounts such as Instagram, Twitter, Garudeya TV Youtube content, and Facebook as an effort to increase public participation. The creation of social media accounts is one of them to reach voters who are currently a millennial generation who are close to technology. Saldana, et al (2015) revealed that one of the factors that encouraged novice voters in political participation was the use of social media, namely the internet. Gallup survey agency (2012) states that one in five people in Indonesia (20.6%) use the internet in their lives, and today, more than half (51%) of Indonesia's young population aged between 15-24 years have used the internet in activities. Most of this young population (96.2%) are social media users.

Furthermore, the results of the global index were quoted from Katadata (2018) of internet users in Indonesia in the age range of 16-64 years. The largest social media platform is Youtube with a percentage of usage of 43%, ranked second on Facebook with a percentage of usage of 41%, then Instagram with a percentage of usage of 38%.

c. Democracy volunteers

Malang City General Election Commission formed Democracy Volunteers. There are 55 (fifty-five) people selected as Malang City who have 10 bases, namely: family base, beginner voter base, young voter base, female voter base, disability base, religious base, marginal base, community base, voter base with special needs, and an internet base.

d. Election smart house

Malang City General Election Commission also provides Election Smart House or *Hamur Garuyeda* in Malang City General Election Commission Office. *Hamur Garudeya* facilitates students to study election activities ranging from miniature elections (miniature polling stations, election simulations, steps to implement voting), how to check the permanent voter list, how to update voter data, and other matters related to election activities. Several times students or the public came to *Hamur Garuyeda* to ask for information about the election or just came to see the latest information. After the simultaneous election event, it was also still used by many parties, for example the guest lecturer of UIN Maulana Malik Ibrahim State Administration Department student on 3 September 2019.

4 CONCLUSIONS

The problems of beginner voters in the 2019 concurrent elections include: novice voters have not done the electronic identity card recording and printing; beginner voters cannot fully use their voting rights even though administratively they can work. The terms of recording, issuance and granting of electronic identity cards can only be done on the day when the population is 17 years old and while if done on election day 17 April 2019 can be considered a violation of regulations; additional voter list issues or select regional moving procedures that are not understood by prospective voters; and the lack of beginner voter understanding of elections.

The level of participation of the people of Malang in the simultaneous elections in 2019 was 82.6%. This exceeds the target of 77.5%. This also includes an increase in voter turnout in Malang. In particular the Malang City General Election Commission has carried out various activities to increase the participation of novice voters, among others, socialization, optimization of General Election Commission social media, the formation of democratic volunteers and providing election smart houses and many more.

REFERENCES

Budiardjo, M. 2008. *Dasar-dasar ilmu politik*. Jakarta: PT Gramedia Pustaka Utama.

Conyers, D. 1994. *Perencanaan sosial di dunia ketiga: Suatu pengantar*. Yogyakarta: Gadjah Mada University Press.

Damsar. 2010. *Pengantar sosiologi politik*. Jakarta: Kencana Prenada Media Group.

Dharmawan, A. T. P. 2016. Inovasi model sosialisasi peran serta masyarakat dalam pemilu. *Jurnal Wacana Politik* 1(2): 12–21.
Irawan, D. 2009. Detik News, 10 Mei 2014, *"Dibanding tahun 2009, angka golput pemilu lebih rendah"*, Retrieved from: (http://news.detik.com/read/2014/05/10 /074125 /2578828 /1562/ dibandingtahun-2009).
Katadata. 2018. *Media yang paling sering digunakan di Indonesia. Tersedia* Retrieved from: https://databoks.katadata.co.id/datapublish/2018/02/01/media-sosial-apayang-paling-sering-digunakan-masyarakat-indonesia.
Mas'oed, M & Collin, M. A. 2008. *Perbandingan sistem politik.* Yogyakarta: Gajah Mada University Press.
Miles, M. B., & Huberman, A. M. (2007). *Analisis data kualitatif: buku sumber tentang metode-metode baru.* Jakarta: UI Press.
Moleong. 2007. *Metodologi penelitian kualitatif,* Bandung: PT Remaja Rosdakarya.
Rapita, D. D. 2014. Dasar pertimbangan hukum putusan hakim Mahkamah Konstitusi terkait pelaksanaan Pileg dan Pilpres Tahun 2014 dan 2019. *Jurnal Arena Hukum UB.*
Republik Indonesia. 1945. Undang-Undang Dasar Negara Republik Indonesia Tahun 1945.
Republik Indonesia. 2008. Undang-Undang Nomor 42 Tahun 2008 tentang Pemilihan Umum Presiden dan Wakil Presiden.
Republik Indonesia. 2011. Undang-Undang Nomor 15 Tahun 2011 tentang Penyelenggara Pemilihan Umum.
Republik Indonesia. 2012. Undang-Undang Nomor 8 Tahun 2012 tentang Pemilihan Umum Anggota DPR, DPD dan DPRD.
Republik Indonesia. (2014). Undang-Undang Nomor 23 Tahun 2014 tentang Pemerintahan Daerah.
Republik Indonesia. 2017. Undang-Undang Nomor 7 Tahun 2017 tentang Pemilihan Umum.
Saldana, M., et al. 2015. Social media as a public space for politics: cross – national comparison of news consumption and participatory behaviors in the United States and the United Kingdom. *International Journal of Communication* 9: 3304–3326.
Sastroatmodjo, S. 1995. *Partisipasi politik.* Semarang: IKIP Semarang Press.
Wardhani, P. S. N. 2018. Partisipasi politik pemilih pemula dalam pemilu. *Jurnal Pendidikan Ilmu-Ilmu Sosial* 10(1): 57–62.

Empowering Civil Society in the Industrial Revolution 4.0 – Sukowiyono et al (Eds)

Legal vacuum of broadcasting supervision on digital social media

Y. Marpi*, Erlangga & R.G. Meru
University of Jakarta, Jakarta, Indonesia

W. Mahardi
University of Indonesia, Jakarta, Indonesia

U. Meylina
University of Lampung, Lampung, Indonesia

ABSTRACT: The study of the monitoring analysis of digital media content broadcasting is deemed necessary as an aspect of legal accountability for digital media content broadcasting based on Law Number 32 of 2002 concerning Broadcasting to ensure legal certainty equal to conventional television broadcasting. With normative juridical research on digital content broadcasting, so far there have been no legal rules that regulate and supervise the implementation and broadcasting so that the emergence of a legal vacuum. Digital conventional broadcasters must comply with the provisions of the broadcasting law to have legal responsibility for broadcasting digital content on YouTube, Net-flix, Instagram, Facebook. The results show that the content created in digital media broadcasting has no legal regulation and provision to obey rules of broadcasting law in the future will have an impact on concerns about the emergence of content that is contrary to the 1945 Constitution and Pancasila.

Keywords: legal vacuum, broadcasting, digital media

1 INTRODUCTION

Television broadcast digitization policy should be viewed as an opportunity to expand and expand the range of various types of broadcasting services for listeners and viewers. This is because so far, the reality of broadcasting carried out by television providers has not reached the audience or listeners evenly with various types of services. Media broadcasting media content YouTube, Netflix, Instagram, Facebook in disseminating the content of broadcast frequencies using the internet. The use of the Internet is based on the amount of space and bandwidth vibration that can only be used by one party, such as concurrent use of space and the amount of vibration and the same width, or share that would interfere (Wicaksono 2020).

Telecommunication frequencies are used to carry or distribute information so that frequency regulation includes: the allocation of frequency bands and designation. Therefore, since the establishment of an international forum in the field of telecommunications, broadcasting in the aspects of technical operations, handled by the ITU so that according to international regulations, broadcasters are part of telecommunications.

In the study of communication, frequency is categorized as public property or public domain. There are three definitions of the frequency status as public domain, namely: (1) public objects, (2) public property, (3) public sphere. At the technical level, broadcast digitization is a term used to describe the process of changing media formats from analog to digital (Mawanda & Muhshi 2019). Technically, digitization is the process of changing all forms of information (numbers, words,

---

*Corresponding Author

DOI 10.1201/9781003180128-2

images, sound, data, and motion) encoded into bits (binary digits) so that data manipulation and transformation are possible including multiplication, subtraction, and addition (Abdullah 1992). The perfection of digital television is expected to ensure the broadcasting industry with an increasingly diverse market. If the previous time broadcast receivers only through broadcasting media content YouTube, Netflix, Instagram, Facebook now with the invention of tablets and smartphones (Adi 2012).

Definition television broadcasting is a communication medium that can be heard by the masses who convey information and ideas openly in public in the form of pictures, video, and sound, which contains a continuous and regular program. While television broadcasting is a communication medium that can be seen and heard by the masses or the general public, which conveys information and ideas openly or privately, which contains a continuous and regular program (Saputra 2020).

Judging from the potential of the technology, broadcasting via optical cable is superior to others, because this technology allows for interactive menus. However, if today the subject of heated discussion is terrestrial broadcasting, it is not independent of the nature of these frequencies that are public domain. The public sphere is the reason for their rules and policies to safeguard the public interest in addition to its power in shaping the information society Law Number 32 of 2002 on Broadcasting (Bachtiar 2015). This public nature also prevents the broadcast migration process from running smoothly according to the predetermined stages. The public still does not understand the implementation of digital broadcasting (Indonesia 2002).

The information socialized contains more about the advantages of digital broadcasting image quality (Lev 2013). Meanwhile, the understanding of this is not well understood. This situation is often exploited by the seller of the television to spend the stock sales of analog TV (Marzuki 2016). While sufficient information on the specifications of televisions compatible with digital broadcasting technology seems deliberately not provided by the seller of the television. Step process of digitalization of television broadcasting in Indonesia is visible in the Regulation of the Minister of Communication and Information Technology Number 2/P/Kominfo/8/2008 on Field Trial Implementation of Digital Television Broadcast (Hakim 2010). Television is only watched without a variety of other activities. Second Hybrid TV namely broadcast simultaneously between TV and the internet. The Internet is expected to be able to bridge with TV.

People's behavior in consuming TV and the internet is different. They consume the internet for more than six hours. While watching TV is only about an hour. Hybrid broadcasting is done to fulfill this kind of behavior so that those who are playing the internet can also watch TV activities. This broadcasting will be able to adapt the various devices in the community to receive TV broadcasts, including mobile TV. Currently, it seems that the Indonesian people are still on Traditional TV. Television as a medium in broadcasting plays a very vital role in contemporary interactions which have urgently been made in a package of regulations. This is because the media have their unique and strong technical capabilities and their compelling contents are the most effective means of information diffusion ever created by humans (Arliman 2017).

The mass media not only crosses geographical boundaries, but also class, racial, cultural, political, educational, and gender boundaries. To distribute the various types of information that are conveyed regularly, both in the form of entertainment that instills and refreshes certain points of view and ways of understanding (Shabana 2018). Hence, the mass media can articulate an ideological synthesis that promotes certain perspectives by excluding others by linking ideological change and sources of authority and helps construct and regulate social reality by structuring some of the most prevalent and most important experiences of their audiences. The era of digital broadcasting has resulted in television broadcasting services that have a function in each frequency channel served by only one television industry.

2 METHODS

This study uses an exploratory research method with an inductive normative juridical approach, namely an approach by drawing general conclusions from real facts in the field. Inductive is a way of thinking, in which general conclusions are drawn from various individual cases. Inductive drawing conclusions begins with compiling general statements. This research was conducted to see the aspects related to the legal vacuum in managing audio-visual recording media broadcasting media content programs YouTube, Netflix, Instagram, Facebook (Wahyuni 2006).

In the process of qualitative analysis, according to Miles & Huberman, namely: data collection, data reduction, data presentation.

3 RESULTS AND DISCUSSION

3.1 *Efforts to overcome the legal void of social media broadcasting*

Regulations are needed to control access to a limited spectrum and ensure that the spectrum is used by people or organizations that are responsible for the public interest. The presence of the internet creates a new paradigm regarding the process of delivering information and messages. In terms of distribution, there are differences in distribution processes that occur due to the shift in conventional media such as television or radio to the new internet media (Abdullah 2020). In conventional television and radio media, content is distributed using terrestrial or cable/satellite waves in one direction, from television stations to the wider community.

By reviewing Law Number 32 of 2002, it turns out that there is no Indonesian Broadcasting Commission (KPI) authority to supervise content on streaming platforms such as Netflix, YouTube, and so on, considering that this media is a digital media that is new and has not many enthusiasts before the formation of the latest broadcast law. So that the first step that will be taken by KPI is KPI plans to revise the Law on Broadcasting (Budiman 2015). This supervision is carried out so that content on digital media is not only worth watching, but also educational and has a moral message that is in accordance with the noble values of Pancasila. Another reason for the purpose of this KPI is because considering that most people have switched from conventional television and radio media and are more inclined to access media or content on streaming platforms.

In conventional television and radio media, content is distributed using terrestrial or cable/satellite waves in one direction, from television stations to the wider community. The emergence of DMBs has influenced future generations of network environments. This socio-technical perspective is based on the theory of social construction which sees technology issues as both technical and social problems (Rachmat 2017).

According to Article 36 of Law Number 32 of 2002, the scope of content that is prohibited and under surveillance is content that is slanderous, incitement, misleading and/or lies, elements of violence, obscenity, gambling, narcotics and drug abuse. Contrasting ethnicity, religion, race and between groups, ridiculing, degrading, harassing, ignoring religious values, Indonesian human dignity, or damaging international relations (Wardah 2014).

To completely eliminate influences, it is not in the sense of free to express opinions, but free of capital, careerism, and bourgeois anarchist individualism. The positive impact of social media broadcasting, such as broadcasting media content on YouTube, Netflix, Instagram, Facebook, is that people can get information quickly, so that the same information message can be received simultaneously and for a moment, communication through the media is considered effective in some conditions.

3.2 *Indonesian broadcasting commission authority in restricting broadcasting of media content on YouTube, Netflix, Instagram, Facebook*

As mandated by Law Number 32 of 2002, KPI was given the authority as well as the function to compile and supervise various regulations related to broadcasting by synergizing broadcasting institutions with the government and society. This arrangement also covers the entire process of broadcasting activities, starting from the establishment stage, operationalization, accountability, and evaluation.

One of the KPIs powers under the Broadcasting Law is to oversee the implementation of broadcasting regulations and code of conduct and broadcast program standards. The aforementioned limitations regarding the objects of the authority of the KPI have been regulated quite clearly in Article 13 of Law Number 32 of 2002. Thus, the authority of KPI as a broadcasting operator is limited to radio and television content, as well as broadcasting institutions (Prabowo 2012). In other words, the KPI is not authorized to supervise new digital media, including You tube, Facebook, Instagram, Netflix, and so on. The issue that there will be an expansion of supervision over the authority of KPI to supervise content in new media, KPI intends to seek regulations that will later become guidelines to add to the object of monitoring digital content from media that has recently been loved by many people such as YouTube, Facebook, Netflix, and the media of a kind.

As referred to by the definition of Broadcasting in Article 6 of Law Number 32 of 2002 concerning Broadcasting, in which broadcasting which is received by the public simultaneously, simultaneously and freely, has a major influence in the formation of opinions, patterns of thought, attitudes and behavior of audiences, so with such an impact, broadcasting organizers should take part in maintaining moral values, morals, culture, personality and national unity based on Pancasila. In line with the task of the KPI institution in supervising content disseminated in the public both on television and radio, as well as content in new digital media, including: YouTube, Netflix, Facebook or similar media provided that they will revise the Law on Broadcasting in terms of KPIs authority to supervise. broadcast content in new digital media. With the discourse of adding KPIs supervisory authority to new media, one of which is Netflix media, the author thinks that this is not urgent to do (Azis 2019).

4 CONCLUSIONS

The emptiness or emptiness of legal science can occur due to various causes besides our tradition of legal discovery which is based on the tradition of jurisprudence which is not included in the flow of legal science so that its development is very slow also due to the rapid progress and growth of community dynamics that cannot be balanced by filling or adding laws. with the tradition jurisprudence that is currently happening. Digital broadcasting technology is a new era in terms of production, distribution, and business. From the production aspect, the supporting equipment for this technology is digital-based. Broadcasters that currently use analog technology, inevitably have to invest in digital equipment. This aspect does not seem to be a big problem in the broadcasting world in Indonesia, because of broadcasting media content on YouTube, Netflix, Instagram, Facebook. The second aspect of the distribution problem relates to the receiving device. Currently, the public does not understand about digital broadcast receiving devices. So that when shopping for a television, this technological aspect is not a consideration. These three business problems. The digital television business model at the center of its industry is the multiplexer implementer. To build a multiplexer requires a large amount of capital. Meanwhile the broadcasters must hire the owner of the multiplexer. For large broadcasters like it is today, investing in digital TV is hardly a problem. From the research activities that have been carried out, the conclusions that the researchers can convey the creative elements of production in the *Pesbukers* program in planning on paper are made not to violate broadcasting ethics. The presentation of shows in the form of dialogue fragments from the players with obscene or insulting words is a form of spontaneity/not found in the script or production briefing before the program activities take place, this is a one-sided improvisation from the players while on stage, the other supporters of the event, namely the studio audience who can interact directly with both verbal context (speech) and attitudes and other players as opposed to dialogue are also able to foster spontaneous attitudes and speech to players who are in dialogue, resulting in a momentary dialogue outside the script.

REFERENCES

Abdullah, A. 2020. The form of broadcasting digitalization policy in Indonesia. *Jurnal Aristo* 08(1): 76–96.
Abdullah, R. 1992. *Hukum acara peradilan tata usaha negara.* Jakarta: Raja Grafindo Persada.

Adi, R. 2012. *Sosiologi hukum: Kajian hukum secara sosiologis*. Jakarta: Yayasan Obor Indonesia.
Arliman, L. 2017. Komisi Penyiaran Indonesia sebagai state auxiliary bodies.*Veritas et Justitia* 3(1): 138–162.
Azis, A.P.A. 2019. Kekosongan hukum acara dan krisis access to justice dalam kasus-kasus pemberhentian kepala daerah/wakil kepala. *Jurnal Hukum & Pembangunan* 49(1): 1–43.
Bachtiar. 2015. *Problematika implementasi putusan Mahkamah Konstitusi pada pengujian UU terhadap UUD*. Jakarta: Penebar Swadaya Grup.
Budiman, A. 2015. Model pengelolaan digitalisasi penyiaran di Indonesia (Model of digitalization of broadcasting in Indonesia). *Politica* 6(2): 107–122.
Hakim, L. 2010. *Kedudukan Hukum Komisi Negara di Indonesia*. Malang: Setara Press.
Lev, D.S. 2013. *Politik dan hukum di Indonesia: Kesinambungan dan perubahan*. Jakarta: LP3ES.
Marzuki, P.M. 2016. *Penelitian hukum*. Jakarta: Kencana.
Mawanda, M.K. & Muhshi, A. 2019. Perlindungan hukum mitra ojek daring di Indonesia. *Lentera Hukum* 6(1): 33–52.
Prabowo, A. 2012. Era penyiaran digital: Pengembangan atau pemberangusan tv lokal dan tv komunitas *Jurnal Komunikasi* 1(4): 301–314.
Rachmat, I. 2017. The use of social media youtube as audio-visual record document management: exploratory study of application the guidelines behavior broadcasting, Standard of Broadcast Programs (P3 and SPS) ANTV on Pesbukers Television Program. *Record and Library Journal* 3(9): 38–49.
Republik of Indonesia. 2002. Undang-undang Nomor 32 Tahun 2002 tentang Penyiaran.
Saputra, A. 2020. *Seret YouTube-Netflix, ini isi gugatan RCTI soal UU Penyiaran, detikNews*. Retrieved from news.detik.com/berita/d-5149091/seret-youtube-netflix-ini-isi-gugatan-rcti-soal-uu-penyiaran.
Shabana, A. 2018. Sistem Stasiun Jaringan (SSJ): Implementasi dan kendala yang dihadapi. *Jurnal Perpektif* 2(2).
Wahyuni, H.I. 2006. Ekonomi politik kebijakan penyiaran Indonesia. *Jurnal Kebijakan dan Administrasi Publik* 2(10): 149–168.
Wardah. 2014. Hak konsumen dalam penyiaran televisi. *Kanun Jurnal Ilmu Hukum* 6(4): 525–537.
Wicaksono, A. 2020. *UU Penyiaran tak berlaku ke Netflix-YouTube, RCTI gugat ke MK, CNN Indonesia*. Retrieved from: https://www.cnnindonesia.com/nasional/20200529181526-12-508073/uu-penyiaran-tak-berlaku-ke-netflix-youtube-rcti-gugat-ke-mk.

Empowering Civil Society in the Industrial Revolution 4.0 – Sukowiyono et al (Eds)

Strengthening the control function of the regional representative council in legislation of the UUD 1945 amendment

A.R.A. Atok*
Universitas Negeri Malang, Malang, Indonesia

ABSTRACT: The UUD 1945 amendment has brought a lot of changes in the Republic of Indonesia's constitutional system. One of them was the establishment of the Regional Representative Council (DPD) as the second chamber of parliament besides the House of Representatives. However, the existence of the DPD in the Republic of Indonesia's constitutional system has not been able to create adequate checks and balances as well as been deemed as half-hearted. The authority and duties given to the DPD by the constitution have not been able to give the DPD control of the legislation. Thus, reviews need to be conducted to give adequate control of the DPD, particularly in the legislation.

Keywords: the control function, legislation, regional representative council

1 INTRODUCTION

The UUD 1945 amendment has presented new institutions, namely the Regional Representatives Council (DPD), which is regulated in article 22C and 22D third amendment of UUD 1945. The presence of DPD in Republic Indonesia's constitutional system is one of the results from the paradigm shift in administering the governance of a centralized to a decentralized government by means of giving the acknowledgment to the regional government to get involved in the determination of public policy. As stated by Manan (2003), the ideas of the establishment of DPD are the proposals to change the representative system to bicameral system and the idea to enhance the involvement in regions on the run of politics and state governance.

The presence of DPD in the Indonesian parliamentary structure of the legislation cannot be separated from the bicameral parliament fundamental ideas. It aims at creating a mechanism which checks between chambers in the parliament by using privileges, among them is the right to veto on the legal product from the other chambers by Jimly Asshiddiqie known as the double check mechanism (Jimly, 2007). For example, the double check on every legislative product to ensure its quality. The presence of DPD as the second chamber in the parliament also ensures the process of legislation in the parliament is not only concentrated in one representative house. Thus, the process of legislation can be done more carefully. The existence of the second chamber is a need in the legislation, as stated by Giovanni Sartori that the concentration of all legislative authority on one institution is not only dangerous but also not wise: two eyes are better than one and the carefulness needs a process of control and limited decision making (Sartori, 1997). The review in this study is aimed at studying how to strengthen the DPD control function in legislation by using conceptual, normative, and comparative approaches.

*Corresponding Author

 DOI 10.1201/9781003180128-3

2 HALF-HEARTED BICAMERAL NO CHECKS AND BALANCES

The existing fact stated that the developed paradigm on the regional representative in the form of DPD is still half-hearted. Thus, it does not give adequate control of the DPD in legislation. This can be seen from the determination of article 22D paragraphs (1), (2), and (3) the third amendment of UUD 1945. Very limited duties and authority were given to the DPD in the legislation. Their duties and authority are limited to (1) proposing certain legislation designs; (2) participating in discussion of certain legislation designs with the House of Representatives (DPR) and President; (3) giving considerations to DPR on certain legislation designs; and (4) supervising on the implementation of certain legislation as well as to report the results to DPR as a consideration for further follow-up.

As stated previously, the very limited authority of DPD on the legislation has implied the nonexistence of a double-checking system. Although institutionally the Indonesian parliament after the amendment of UUD 1945 consisted of two chambers, the nonexistence of a mechanism of checking on each other between both chambers then has several advantages of a bicameral system stated by some experts will not be realized.

The relationship pattern of DPR and DPD in the legislation in the aforementioned half-heartedly bicameral cannot fulfill the political justification and efficiency as stated by Tsebelis and Jeannette (1977). The justification for the bicameral system can be seen from the political and efficiency aspects. from the political aspect, bicameral system: (a) can prevent the majority tyranny, since the existence of a veto right given institutionally; (b) can prevent minority Tyranny, since both chambers require support from a broad constituency to support the legislation; (c) can decrease corruption and inhibit a corrupt legislation process, since the possibility of collusion implementation in the bicameral system is more difficult compared to the unicameral system; (d) can decrease costs of decision making, due to the existence of the conference committee mechanism.

The relationship pattern of DPR and DPD in the legislation as determined in the amendment of UUD 1945 and other legislation from political aspects has not been able to prevent the possibility of majority tyranny incidence. This happens because DPD does not have a veto right to file an objection or the delay given institutionally in the determination of a legislation agreement. In terms of efficiency, it still cannot assure to generate a better and more stable legislation. It has not been able to create the possibility of quality control and cannot decrease the possibility of corruption and collusion of the DPR and government. It also cannot reduce the costs of decision making due to the nonexistence of conference mechanisms.

The bicameral system in Indonesian parliament after the amendment of UUD 1945 does not create checks and balances. As stated by Tsebelis and Jeannette (1977), about the cases when the highest assembly only has a consultative authority and lower authority in the final decision making. Thus, such bicameral does not create the checks and balances system.

In fact, conceptually, the parliament system after the amendment of UUD 1945 cannot be stated as implementing a bicameral parliament even in the weak category. This is because of the results of the compromise between bicameralism and unicameralism, thus creating a half-hearted bicameral (Gaffar, 2003).

3 BETWEEN WEAK BICAMERALISM AND MEDIUM-STRENGTH BICAMERALISM

When seen from the aspects of very limited and unequal duties and authorities compared to DPR in the legislation then the parliamentary system is followed after the amendment of UUD 1945. Referring to the parliamentary system theory as stated by Lijphart (1984), it can be categorized into a weak bicameralism with asymmetric authorities. However, when viewed in terms of the selection method, it cannot be categorized as weak, because even though the membership structure of the two chambers is incongruent, the selection method used to select members is the same, namely through general elections. In fact, the selection method of elections that is used to elect DPD members that use the most representative district systems with individual participants is seen as more democratic than those used to elect DPR members that use an open proportional system with participants from political parties. Thus, the legitimacy of the DPD is not weaker or even more powerful than the legitimacy of the DPR. Therefore, as seen from the selection method aspect and the structure of the membership it can be categorized as medium-strength bicameralism. Such a pattern can be said as deviated from the general pattern of bicameralism.

In the strong bicameralism parliament, the legislation usually begins from any chamber and needs a consideration by both assemblies in the forum which are not yet confirmed. Meanwhile, in the weak bicameralism, one chamber has a higher status than another (Purnomowati, 2005). The existence of the DPD as the second chamber in the Indonesian parliament after the UUD 1945 amendment can be categorized as weak, it is because the discussion process of the legislation can only be done in the DPR as the first chamber which has the stronger position, while the DPD as the second chamber only has a right to propose to participate in the discussion (not independent discussion) and gives a consideration on certain legislations designs without having a veto right. According to Purnomowati (2005), the position of DPD as such is rated as "too weak".

Compared to other nations which also have a weak bicameral parliament, the bicameral parliament structures of DPR and DPD can be seen from the comparison of formal authorities owned by both and can be stated as "too weak." As in Algeria, the People's National Assembly as the first chamber, the members are elected publicly, directly, and in secret. Meanwhile, as the Council of Nation as the second chamber, the number of the members is 1/2 of the number of the People's National Assembly. They were elected by the People's Communal Assemblies. 2/3 of the members were elected by the People's Wilaya Assembly, and 1/3 of the members were assigned by the president (Article 101 Constitution of the People's Democratic Republic of Algeria).

Although the Council of the Nation does not have the authority to propose legislations designs, it has the same right as the People's National Assembly to give an agreement on the legislation's designs. Council of the Nation as the second chamber has more function as the control unit on the proposed legislation design and has been approved by both the People's National Assembly and government. There is no legislation design which can be legitimate to be constitutions without the Council National approval. When a deadlock occurs in the achievement of agreement with People's Nation Assembly and Council of the Nation, then the government proposes a joint committee to be formed into both chambers with the same number of members to do a joint discussion to combine opinions so that mutual agreement can be obtained (Article 119 and 120 Constitution of the People's Democratic Republic of Algeria).

In addition, France as a unitary nation with a weak bicameral system has the National Assembly as the first chamber with the members elected directly, while Senate as the second chamber, its members were elected indirectly to represent each republican region unit (Article 24 Constitution of France Republic). The Senate as a second chamber, in general, is a subordinate of the National Assembly but has a somewhat larger role in influencing organic law and financial bills even though the decision remains at the National Assembly as the first chamber. Similar to the National Assembly, the Senate as the second room also has the authority to accept the initial proposal of legislation designs from the Government, unless the Government's initiative legislation design on Finance and Social Security must be submitted first to the National Assembly. However, the legislation design on a regional organization and about institutions representing the French nation abroad must first be submitted to the Senate. Both the National Assembly and Senate both have the authority to discuss the legislation design from the Government submitted to him. As a control and balance, the same authority is given to the National Assembly and Senate to give consideration to the legislation design that has been approved by one of the chambers. The dissent between the National Assembly and Senate is resolved by a joint committee formed by the prime minister, if the joint committee cannot achieve an agreement, then the authority is given more to the National Assembly for decision making (Article 39 – 47 Constitution of France Republic).

From this comparison, it is significantly apparent that the DPD as the second chamber in the bicameral parliament system in Indonesia tends to have characteristics of being conservative, very weak, and very dependent on the DPR's political interests. DPD does not have an adequate role in implementing the checks and balances function of the legislation. Conceptually, the mechanism of checks and balances does not only apply to legislative and executive units but also occurs and is institutionalized between chambers and representative councils in the legislative body.

Fundamentally, the bicameral parliament concept is constructed to encourage the existence of checks and balances with the legislative body. Therefore, in the bicameral parliament, there needs to be a competition between both chambers to trigger the condition of mutual control which enables the balance in the parliament. This is also according to the consideration of both chambers in the parliament in which in general represent different constituents.

The improvement on the control function of DPD in the legislation should not be directed to strong bicameralism or even perfect bicameralism, but it is enough in the medium bicameralism by giving adequate authorities to implement the control function. The use of the wordings strong and perfect bicameralism can cause difficulties, so that the federal state with a parliamentary system of government also in practice uses weak bicameralism by giving an authority to the second chamber that is not as strong as the first chamber and limiting the authority on the second chamber to only have a veto power over certain legislation designs. However, there are also unitary states which implement a strong bicameralism such as Columbia and Algeria.

The use of the wordings strong or even perfect bicameralism for Indonesian parliament has been proposed and submitted to the DPD as the material for the Fifth Amendment of UUD 1945 (Kelompok DPD di MPR RI, 2011). In the discussion of the UUD 1945 amendments in MPR (People's Consultative Assembly) particularly on the second and third amendments, the proposal has long been debatable (Sekjend Kepaniteraan & Mahkamah Konstitusi, 2008). Although the proposal of the usage of strong bicameralism was also supported by some constitutional law experts, in which in the end MPR stipulates the use of a "half-hearted" bicameral system that is too soft. There appear to be concerns from some of the MPR factions if the Indonesian parliament are made a strong bicameral system, feared that the structure of the state organization will develop towards Federalism (Sekjen MPR RI, 2001). Actually, these concerns do not need to occur because research results by Lijphart (1999) revealed that 13 out of 36 nations with unitary forms use a bicameralism system.

It is also in line with Fatmawati (2010) research which shows that from 22 nations which use a bicameralism system, 10 of them are unitary nations and 12 of them are federal nations. However, 2 out of 10 nations use a strong bicameralism, 7 nations use Medium-Strength Bicameralism or Weak Bicameralism. In addition, 2 out of 22 nations in which both chambers do not have the authorities of the legislation, 2 nations with both chambers do not have equal authorities as the first chamber (perfect bicameralism), 7 nations with both chambers have authorities in the legislation almost equal to the first chamber (strong bicameralism), and 11 nations with both chambers do not have equal authorities to the first chamber (weak bicameralism).

To strengthen the control function of in the legislation, according to the researcher, the DPD needs to be given some authority to participate in giving the joint agreement with the DPR and the president on the discussion results of legislation design. The DPD also needs to be given the authority to participate in the joint discussions with the DPR and president even without the right on decision making. This will enable DPD to give an argument and also escort the delivered consideration. This is also meant to open an opportunity and also encourage the control on each other between DPR and the president of responding to the DPD's consideration.

4 CONCLUSIONS

In the process of legislation, the DPD indeed should not necessarily be given an equal authority on the House of Representatives (DPR) or President. However, the DPD needs to be given the authority that makes it possible to do checks and balances in order for DPD participation to get a better quality of the legislation. Even though the authority is quantitatively limited, when the process of the participation is qualified, it will be more meaningful compared to a broad scope but the participation is counterfeit.

REFERENCES

Algeria Constitutional Background. 1996. Retrieved from: *http://www.oefre. unibe.ch/law/ic/ag.html.*

Ashiddiqie, J. 2007. *Hukum tata negara darurat.* Jakarta: PT. Raja Grafindo Persada.

Constitution of France Republic. 2008. Retrieved from: *http://www.oefre. unibe.ch/law/icl/.*

Fatmawati. 2010. *Struktur dan Fungsi Legislasi Parlemen dengan Sistem Multikameral, Studi Perbandingan antara Indonesia dan Berbagai Negara.* Jakarta: UI Press.

Gaffar, J. M. et. al. (ed). 2003. *Dewan Perwakilan daerah dalam sistem ketatanegaraan Republik Indonesia.* Jakarta: Kerjasama Antara Sekretariat Jenderal MPR RI dengan UNDP.

Kelompok DPD di MPR RI. 2011. *Naskah perubahan kelima Undang-Undang Dasar Negara Republik Indonesia Tahun 1945 Usul Perubahan Pasal Beserta Alasannya.*

Lijphart, A. 1984. *Democracies pattern of majoritarian and consensus government in twenty-one countries.* New Haven And London: Yale University Press.

Lijphart, A. 1999. *Pattern of democracy: Government form and performances in thirty-six countries.* New Haven and London: Yale University.

Manan, B. 2003. *DPR, DPD, dan MPR dalam UUD 1945 Baru.* Yogyakarta: FH UII Press.

Purnomowati, R. D. 2005. *Implementasi Sistem Bikameral dalam Parlemen Indonesia.* Jakarta: Rajawali.

Republik Indonesia. 1959. *Undang-Undang Dasar Negara Republik Indonesia Tahun 1945.* Lembaran Negara Republik Indonesia Tahun 1959 Nomor 75.

Republik Indonesia. 1999. *Perubahan Pertama Undang-Undang Dasar Negara Republik Indonesia Tahun 1945.* Jakarta: Sekretariat Jenderal MPR RI.

Republik Indonesia. 2000. *Perubahan Kedua Undang-Undang Dasar Negara Republik Indonesia Tahun 1945.* Jakarta: Sekretariat Jenderal MPR RI.

Republik Indonesia. 2001. *Perubahan Ketiga Undang-Undang Dasar Negara Republik Indonesia Tahun 1945.* Jakarta: Sekretariat Jenderal MPR RI.

Republik Indonesia. 2002. *Perubahan Keempat Undang-Undang Dasar Negara Republik Indonesia Tahun 1945.* Jakarta: Sekretariat Jenderal MPR RI.

Republik Indonesia. 2014.*UU No. 17 Tahun 2014 tentang MPR, DPR, DPD, dan DPRD.* Lembaran Negara Tahun 2014 Nomor 182 Tambahan Lembaran Negara Republik Indonesia Nomor 5568.

Republik Indonesia. 2017. *UU No. 7 Tahun 2017 tentang Pemilihan Umum.* Lembaran Negara RI Tahun 2017 Nomor 182, Tambahan Lembaran Negara RI Nomor 6109.

Republik Indonesia. 2014. *UU No. 42 Tahun 2014 tentang Perubahan atas UU No.17 Tahun 2014 tentang MPR, DPR, DPD, dan DPRD.* Lembaran Negara Republik Indonesia Tahun 2014 Nomor 383, Tambahan Lembaran Negara Republik Indonesia Nomor 5650.

Republik Indonesia. 2018. *UU No.2 Tahun 2018 tentang Perubahan Kedua atas UU No.17 Tahun 2014 tentang MPR, DPR, DPD, dan DPRD.* Lembaran Negara RI Tahun 2018 Nomor 29, Tambahan Lembaran Negara RI Nomor 6187.

Sartori, G. 1997. *Comparative constitutional engineering an inquiry into structures, incentives and outcomes.* New York: New York University Press.

Sekretariat Jenderal MPR RI. 2001. *Buku Kedua Jilid 3A Risalah Rapat Panitia Ad Hoc I badan Pekerja MPR RI ke 11 s.d ke 15 Tanggal 20 Maret s.d 15 Mei 2001 Masa Sidang Tahunan MPR RI Tahun 2001.* Jakarta: Sekretariat Jenderal MPR RI.

Tim Penyusun Naskah Komprehensif Proses dan Hasil Perubahan UUD 1945. 2008. *Naskah Komprehensif-Perubahan Undang-Undang Dasar Negara Republik Indonesia Tahun 1945, Latar Belakang, Proses dan Hasil Perubahan 1999-2002 BUKU III Lembaga Permusyawaratan dan Perwakilan Jilid 1 dan Jilid 2.* Jakarta: Sekretariat Jenderal Kepaniteraan dan Mahkamah Konstitusi.

Tsebelis, G., & Jeannette, M. 1977. *Bicameralism.* London: Cambridge University Press.

Empowering Civil Society in the Industrial Revolution 4.0 – Sukowiyono et al (Eds)

The procedure of execution on foreign arbitral awards in Indonesia and issues

S.B. Yunari*
Universitas Tri Sakti, Jakarta, Indonesia

ABSTRACT: This paper aims to discuss regulations, procedures and problems arising from the implementation of foreign arbitration awards in Indonesia. This study uses legal research with secondary legal materials obtained from literature research for primary, secondary and tertiary legal materials to be analyzed prescriptively against the laws and regulations governing the implementation of foreign arbitration in Indonesia. One of the important issues in dispute resolution through foreign arbitration is the existence of obstacles in the implementation of the foreign arbitration award requested by the exequatur through the Central Jakarta District Court, Indonesia. The constraints found are more directed at not fulfilling the formal requirements of the exequatur and technical problems when the execution of the decision will be carried out by the clerk of court in practice when the foreign arbitration award in question has been accepted for implementation in Indonesia.

Keywords: arbitral-awards, foreign, execution, procedure, Indonesia

1 INTRODUCTION

Asian Economic Community (referred as AEC) implemented on ASEAN countries is intended to establish a single market cycle in the Southeast Asia region. ASEAN leaders agree that AEC will form a single market in the Southeast Asia region at the end of 2015 (ASEAN 2025). The establishment of this single market is believed to enable a country in marketing its "goods and services" to other countries throughout Southeast Asia in the midst of stringent competition, because AEC does not merely open the flow of goods and service commerce, but also market for professionals, such as lawyers, doctors, accountants, and other concerned parties. If the emerging competition podium has been established, particularly in the industry and human resources sectors such as medical experts, the competition in the trade and industry sectors will potentially cause dispute, such as in the implementation of contracts or business cooperation between each country.

Hence, the existence of AEC does not merely affect the economy sector but also the legal sector. With the implementation of AEC in 2015 on ASEAN countries, the legal sector deals with the existence of institutions for dispute settlement in international trade transactions appointed by member countries of AEC. Foreign arbitration institution as one of the dispute settlement institutions has been mostly preferred by countries that should play a bigger role, if chosen as a dispute settlement institution by member countries of AEC, including Indonesia. Therefore, It is required for the concerned parties to give attention to goods and services to be "recognized and enforced" in Indonesia.

Recently, there are several reasons why business people employ arbitration as dispute settlement institution in business transaction, such as: the existence of freedom and extensive autonomy for the parties, the confidence and security that it gives to situation, uncertainty in connection with

*Corresponding Author

different legal system governing the parties; the parties have confident in the arbitrator who has expertise in dispute settlement; arbitration is seen as faster way in rendering the award which is less expensive hence cost-effective; in rendering the award the parties are protected from matters that may jeopardize their business interest in public. Arbitration applies values that are naturally private in rendering the award and waive public matter; enforcement of award is more convenient.

Based on the aforementioned reasons, it is undeniable that arbitration is exceptionally attractive for business people as a dispute settlement institution, particularly in the international business environment. Currently, there are several foreign arbitration boards as dispute settlement institutions whose existence is recognized by ASEAN countries and worldwide countries that are members of international organizations, such as: "The International Chamber of Commerce, The International Centre for The Settlement of Investment Dispute and United Nations Commission on International Trade Law". Another important matter that also requires attention in settlement of international business dispute is the enforcement of international arbitral-award by the disputing countries, because a dispute already decided by foreign arbitration board will not be effective if pointless to be enforced in a country for any reason.

Hence, the parties engaging an international business transaction within ASEAN or in global scope, must find out the commitment of the business partner country; whether it has ratified the treaty on "the recognition and enforcement of foreign arbitral awards" issued by International organization of United Nations such as the New York Convention of 1958 concerning about Recognition and Enforcement of Foreign Arbitral-Awards (referred "New York Convention"), that has come into effect as of June 7, 1959. Currently, New York Convention has been ratified by 150 countries, including Indonesia, declaring its ratification and accession of New York Convention by Presidential Decree Number 34/1981 on "The Ratification of the Convention on The Recognition and Enforcement of Foreign Arbitral-Awards" on August 5, 1981 (referred as Presidential Decree Number 34/1981) (Claver-Carone 2002; Kadir & Murray 2019).

A country ratifying New York Convention such as Indonesia must recognize and enforce any applied foreign arbitral-award for recognition and enforcement by countries that fulfill the requirements as provided in Articles 65–69 of Law Number 30/1999 on Arbitration and Alternative Dispute Settlement (referred as Law Number 30/1999) (Isrihadi 2018). Practically, it provides guidelines on how foreign arbitral-awards are enforced in Indonesia, including the obstacles related to foreign arbitral-awards. However, it remains debatable whether foreign arbitral-awards are nullified or challenged. Thus, this paper aims to address such issues by presenting understanding and illustration to academic community and legal practitioners in Association Southeast Asian Nations in order to understand law implementation of foreign arbitral-awards in Indonesia (Hamzah 2018).

2 METHODS

The type of this research is "legal research" or "doctrinal research" with secondary legal material obtained from library research for primary, secondary and tertiary legal material to be prescriptively analysed on laws and regulations governing arbitration as well as relevant sectoral laws. The methods in this research are statute and cases approach which are conducted by analyzing all the concepts related to the legal issues of foreign arbitration (Hutchinson 2002; Marzuki 2011; Yaqin 2007)

3 RESULTS AND DISCUSSION

3.1 *Definition and form of foreign arbitration*

As foreign arbitration commonly recognized as "international arbitration" or "commercial arbitration" is a term employed by New York Convention, where foreign Arbitration definition under it becomes an institution/person that settles dispute among citizens, issued in a state other than the state where the award is recognized and enforced (article 1) (Fry 2009; Girsang 1989). The form of foreign arbitration institution/body is Ad-hoc or institutional (Fauziah 2009). For example, the example national institutional international arbitration board in Indonesia is Indonesian National Arbitration Agency (BANI) (Butler 2009). In general, each country has a nationally Institutional Arbitration as center of national arbitration in the respective country, such as: Nederland's Arbitrate Institute, The Japan Commercial Arbitration Association, The American Arbitration Association, The British Institute of Arbitrators etcetera. In addition to the regional Institutional Arbitration established by third world countries (developing countries), such as *Asia Afrika Legal Consultative Committee* (AALCC), the International Chamber of Commerce (ICC), Investment Dispute (ICSID), United Nations Commission on International Trade Law (UNCITRAL) (Asouzu 1999; Efrat 2016; Yunari 2017).

3.2 *Scope of recognition and enforcement of foreign arbitral-awards*

In principle, awards are rendered by foreign arbitration board, ad-hoc as well as institutional, to countries that have ratified the New York Convention, and must be recognized and enforced. It is in line with Article 1 paragraph (1) of New York Convention that states foreign arbitral-awards: "arbitral awards are made in the territory of a state other than the state where the recognition and enforcement of such awards are sought" (Amro 2013).

Hence, if discussing foreign arbitral-awards, the issue will regard the scope of requirements in enforcing such this institution in the District Court of Central Jakarta as regulated in Article 65, Law Number 30/1999. The requirements regulate that a foreign arbitral-award are "eligible to be recognized and enforced" in Indonesia which include:

1) International arbitral-awards made by arbitrator or council of arbitrators in a state that is bound by an agreement with Indonesia, bilaterally or multilaterally, on the recognition and enforcement of international arbitral-awards" (*reciprocity principle*).
2) The relevant international arbitral-awards are limited to awards within the scope of the law of commerce under Indonesian laws (Mertokusumo 2013).
3) International arbitral-awards that can be enforced in Indonesia are limited to those awards that are not contrary to public order.
4) Foreign arbitral-awards can be enforced in Indonesia, after obtaining an enforcement order (*exequatur*) from the Head of the District Court of Central Jakarta.
5) Foreign arbitral-awards that involve "The Republic of Indonesia" as a party to a dispute can only be enforced after obtaining *exequatur* from the Supreme Court of the Republic of Indonesia, and thereafter reverted to the District Court of Central Jakarta.

3.3 *Procedures of execution on foreign arbitral-awards in Indonesia*

The procedures to obtain *exequatur* on foreign arbitral award in Indonesia through Central Jakarta District Court are as follows:

1) The application for enforcement of foreign arbitral award is made after the relevant award is submitted and registered by the board of arbitration/arbitrators residing outside the territory of Indonesia, represented by the Arbitrator or legal representative of the Applicant (Indonesian advocate) through Junior Clerk of Civil Affairs at Secretary Clerk of District Court of Central Jakarta.
2) Submission of enforcement application documents must be accompanied with:
   a) Certificate of registration/deponir of the applicant;
   b) Special Power of Attorney, if the request is performed by the attorney deponir applicant;
   c) *Exequatur* requests a letter from Clerk Secretary of District Court of Central Jakarta;
   d) Original or certified authentic copy of the foreign arbitral award and its formal Indonesian translation.
   e) Original or certified authentic copy of agreement based on which the foreign arbitral award is made and its formal Indonesian translation.

f) Originally a statement from "the diplomatic representative of the Republic of Indonesia" in the country where the foreign arbitral award is made stating, that the state of applicant is bound by "bilateral or multilateral agreement" with the Republic of Indonesia concerning the recognition and enforcement of foreign arbitral awards.

3) After obtaining registration (*deponir*) number, at the latest 1 (one) week as of the submission of application for enforcement documents of such foreign arbitral award, the enforcement officer/bailiff will examine the requirement of application for *exequatur* that has been submitted, and then the bailiff will prepare a resume stating the nature of the foreign arbitral award as sought by *exequatur*, whether it is a condemnation award or declaratoire award, to have the opinion of Head of the Central Jakarta District Court.
4) Head of the Central Jakarta District Court will provide opinions related to the legal considerations of whether the foreign arbitral award is "acceptable" or "rejected". If acceptable, it means that "the foreign arbitral-award can be recognized and enforced in Indonesia", and subsequently the Head of the Central Jakarta District Court will issue an order (Exequatur) that the foreign arbitral-award is "eligible to be enforced". However, if refused, then the foreign arbitral-award can be submitted for classification to the Supreme Court. Within a maximum of 90 (ninety) days as receipt of the classification request, the Supreme Court shall provide consideration and decide the request.
5) As a follow up of foreign arbitral-award that has been accepted by the Head of the Central Jakarta District Court after obtaining *exequatur*, the bailiff will notify the respondent in the form of summons (*aanmaning*), by serving notice to the parties to perform the foreign arbitral-award, at the latest 8 (eight) days as of the summon (*aanmaning*). Thereafter, enforcement attachment can be performed with the assistance of the bailiff, on assets and property of the respondent for enforcement. The procedures of attachment, and enforcement of foreign arbitral-award against the respondent for enforcement shall follow the procedures in civil law.
6) If the respondent domiciles outside the jurisdiction of the Central Jakarta District Court, then the serve of notice will be made through delegation by requesting the assistance of the District Court where the respondent domiciles.
7) If the parties have performed the award, then the applicant and the respondent shall submit a notification accompanied with proof that the parties have performed such foreign arbitral-award.

Based on the notification and proof, the enforcement section will discharge the registered foreign arbitral-award case from the registration number that has been previously recorded.

3.4 *Issues in the implementation of foreign arbitral-awards in Indonesia*

Based on research, there were 92 cases filed by petition "*deponir*" and "*exequatur*" in the District Court of Central Jakarta; consisting of 52 cases which constituted a request deponir and 41 cases were plea *Exequatur* and entirely granted by Head of the Central Jakarta District Court, to be later proceeded with the *meaning* of the parties by the Central Jakarta District Court bailiff. However, there are still constraints in the execution of foreign arbitral-awards which requested *deponir* and *exequatur* in the Central Jakarta District Court, because:

1) There is no clear standard fee for filing registration (*deponir*) and *exequatur* of foreign arbitral-awards in the Central Jakarta District Court. Therefore, *deponir* application and *exequatur* are not similar to the registration and application filing in the Central Jakarta District Court. This finding obviously needs to be regulated by the Central Jakarta District Court, if certain costs incurred related to legal certainty and *Exequatur* in the Central Jakarta District Court. This reason is in contrast with the arrangement before the enactment of Law Number 30/1999, in which the first *Exequatur* costs carried by the clerk of the Secretary of the Supreme Court was IDR. 250.000 – (two hundred and fifty thousand rupiah), deposited through "Bank BRI Branch Office of Veterans".
2) The absence of clear and strict rules regarding the status of the continuation of Registration (*deponir*) or *Exequatur* was inevitable. In relation with petition *deponir* course, after application

registration is completed, then there is no clear mechanism to follow up deponir by the parties, if a foreign award has been executed by the parties. Similarly, if the application has been given by Head *Exequatur* of the District Court of Central Jakarta, there is no clear and firm mechanism; therefore, foreign arbitral-awards requesting *exequatur*, will otherwise be executed by the parties, with the proven existence proof of the statement from the parties completing the decision, and is submitted in writing to the District Court of Central Jakarta bailiff.

3) With regard to jurisdiction, if the execution of foreign arbitral-awards are outside the territory of the District Court of Central Jakarta, a foreign award shall be conducted by the District Court in another Indonesian region, then it becomes a problem because there is no clear and unequivocal rule, to whom the cost of the delegation and the calling of the parties shall be borne.
4) The limitations shall be emphasized on "requirements of foreign arbitral-award", which must not conflict with the "public order" in Indonesia, to prevent different interpretations in practice.

4 CONCLUSIONS

Based on the above discussion, this research concludes that Indonesia has principally set regulations on the recognition and enforcement of foreign arbitral-awards in the Presidential Decree Number 34/1981 on the Ratification the New York Convention, 1958, as well as procedures for the implementation of foreign arbitral awards. Exequatur is regulated in Article 65–69 of Law Number 30/1999. According to Law Number 30/1999, the only one institution authorized for "the filing of the application for Registration application (depone)", and "Exequatur for the recognition and enforcement of foreign arbitral-awards" in Indonesia is the District Court of Central Jakarta (Relative Competence). As an exception, if the disputing parties are not in the State of Indonesia, then an exequatur will be issued by the Supreme-Court. Upon obtaining an exequatur by the Supreme-Court, then the exequatur documents will be reverted to the District Court of Central Jakarta for the execution. Foreign arbitral awards which recognition and enforcement have been accepted by the Head of District Court are ineligible to be appealed nor submitted for classification. However, foreign arbitral awards whose recognition and enforcement have been rejected by the Head of District Court, are eligible to be submitted for classification. To add, the Supreme Court decision remains unchallenged.

REFERENCES

Amro, I.A.S. 2013. *Recognition and enforcement of foreign arbitral awards in theory and in practice: a comparative study in common law and civil law countries*. Cambridge: Cambridge Scholars Publisher, ProQuest Ebook Central.

ASEAN. 2015. *Forging ahead together*. Jakarta: ASEAN Secretariat.

Asouzu, A.A. 1999. African states and the enforcement of arbitral awards: Some key issues. *Arbitration International* 15(1): 1–52.

Butler, N. 2019. Non-disputing party participation in ICSID Disputes: Faux Amici? *Netherlands International Law Review* 66(1):143–178.

Claver-Carone, M.J. 2002. Post-handover recognition and enforcement of arbitral awards between Mainland China and Hong Kong SAR: 1999 agreement vs. New York convention. *Law and Policy in International Business* 33(2): 369–403.

Efrat, A. 2016. Promoting trade through private law: Explaining international legal harmonization. *The Review of International Organizations* 11(3): 311–336.

Fauziah, F. 2019. Annuitant legal protection associated with Act Number 40 of 2014 About Insurance Business. *Journal of Legal, Ethical and Regulatory Issues* 22(1): 1–7.

Fry, J.D. 2009. Disorder public international under the New York Convention: Wither truly international public policy. *Chinese Journal of International Law* 8(1): 81–134.

Girsang. S.U.T. 1989. *Arbitrase*. Jakarta: Publisher Djambatan.

Hamzah. 2018. Bilateral investment treaties (BITS) in Indonesia: A paradigm shift, issues and challenges. *Journal of Legal, Ethical and Regulatory Issues* 21(1): 1–13.

Hutchinson, T. 2002. *Researching and writing in law.* New South Wales: Published Lawbook Co.
Isrihadi, E.I. 2018. A study of commercial arbitration and the autonomy of the Indonesian arbitration law. *Journal Legal, Ethical and Regulatory Issues* 21(1): 1–8.
Kadir, M.A. & Murray, A. 2019. Resource nationalism in the law and policies of Indonesia: A contest of state, foreign investors, and indigenous peoples. *Asian Journal of International Law* 9(2): 298–333.
Marzuki, P.M. 2011. *Legal research.* Jakarta: Publisher Kencana Prenada Media Group.
Mertokusumo, S. 2013. *Civil law of procedure.* Yogyakarta: Publisher Cahaya Atma Pustaka.
Yaqin, A. 2007. *Legal research and writing.* Malaysia: Malayan Law Journal Sdn Bhd-LexisNexis.
Yunari, S.B. 2017. *Effectiveness of enforcement of foreign arbitral awards in Indonesia.* Result of Research Report. Dept. of Business Law, Faculty of Law, Universitas Trisakti, Jakarta.

Empowering Civil Society in the Industrial Revolution 4.0 – Sukowiyono et al (Eds)

Law analysis of birth registration based on the biological father's recognition

E.P. Handayani, Z. Arifin*, S. Firdaus & L. Dwijayanto
Kadiri Islamic University, Kediri, Indonesia

ABSTRACT: This study aimed to analyze the importance of the biological father's recognition of children in birth registration reviewed from civil law and the procedure of a biological father recognizing a natural child that had been recorded in the civil registry administratively based on the marriage certificate. The problem formulation in this study was answered with normative legal research. The results showed that the recognition of biological fathers to children in birth registration was very important because it was very influential in the life of the child today and in the future. The procedure conducted by the biological father in acknowledging the biological child that had been recorded on the civil registry in the sequence was that the applicant filed a lawsuit to the court with the conditions that have been determined, prepared evidence in the form of DNA test results, and proposed two witnesses to provide information and strengthen the lawsuit.

Keywords: natural child, recognition, birth certificate

1 INTRODUCTION

The presence of a child in a normal marriage is one of the dreams for every married couple. But in reality, there are some people who throw away their own children to cover up disgrace for their families. The disposal was carried out because the child born was an out-of-wedlock child who was not justified in the religious and ethical tenets that prevail in society (Witanto, 2012). The status of the child in family law is grouped into two kinds, namely legal children and natural children (an out-of-wedlock child). The legality is stipulated in Law Number 1 of 1974 Article 42 which reads "A legitimate child is a child born in or as a result of a lawful marriage." A legitimate child has a perfect civil relationship with his parents (Afandi, 1982).

The civil relations in question include the right of *nasab* (lineage) of children connected to the father (in Islam), the right to fulfill the living from parents to children, the right to care and education, the right to inherit each other, the right of marriage for the father of daughters, and other civil rights (Hadikusuma, 2013). The natural child is a child born outside of a legal marriage between his parents.

An illegitimate child has only a civil relationship with his mother and his mother's family. The child has no legal relationship with his father because legally both religious law and national law he has no blood relation with the man who is his biological father. The child out-of-wedlock does not obtain the material and moral rights that should be obtained by a child from his father, such as the right to care, the right to a living, the right to marry for the daughter, and the right to inherit from each other in the event of death.

*Corresponding Author

DOI 10.1201/9781003180128-5

In principle from a biological point of view, it is impossible for a child not to have a father. Therefore, for the sake of the law concerning all consequences in the field of inheritance, citizenship, guardianship, and so on, there needs to be recognition and ratification of children so as to create a new civil legal relationship (Situmorang & Sitanggang, 1991). There are several studies that have studied the problem, among others: *Critical Study on Child Recognition Institute (Critical Study Between Islamic Law and Positive Law),* conducted by Ahmad Darbi B; *The Court's Viewpoints on the Decision of Child Foster Proposals* conducted by Sirajuddin S; and *The Position of Natural Child: Analysis of the Decision of the Constitutional Court Number 46/PUUVIII/2010 concerning Material Test of Article 43 Paragraph 1 of Law Number 1 Year 1974 concerning Marriage* by Muhamad Arifin. However, the study has not examined the importance of child recognition by biological fathers in depth. Therefore, this study aimed to analyze the importance of the biological father's recognition of children in the recording of birth and the procedure of a biological father to recognize a biological child that had been recorded in the civil registry administratively based on the marriage book.

2 METHODS

To answer research questions, this study used normative legal research methods. Normative juridical methods were carried out by document studies to study existing norms or laws and regulations related to the status and rights of children and recognition of children (Soekanto & Mamudji, 2004; Marzuki, 2011). The primary legal sources used are the Constitution of the Republic of Indonesia Year 1945, Law Number 1 of 1974 concerning Marriage, Law Number 23 of 2002 concerning Child Protection, the Book of Criminal Law, a Compilation of Islamic Law, and the Book of Civil Law. Secondary sources include official documents, books, research results in the form of reports, journals, opinions, and legal opinions.

3 RESULTS AND DISCUSSION

3.1 *The importance of biological father recognition for the child*

Based on the decision of the Constitutional Court Article 43 paragraph (1) of the Marriage Law that stated, "The natural child has a civil relationship with his mother and his mother's family and with the man as his father which can be proven based on science and technology and/or other evidence according to the law has a blood relationship including civil relations with his father's family." So, the natural child has only a civil relationship with his mother and his mother's family only.

The natural child has no relationship with his father so has no rights and obligations to his father, therefore the natural child has legally required recognition to create a civil relationship between a natural child and the father as a parent. Article 280 of the Civil Law stated that "By committing an act of confession to a natural child, a civil relationship has emerged between the child, and his father or mother".

According to the applicable civil law in Indonesia, the determination of a child's origin can be made by voluntary recognition and implemented recognition. Voluntary recognition of a child is a statement as determined in civil law that a mother and father or mother admit a child born to a mother is a true child from the result of their biological relationship and that the relationship is not in a valid marital bond, and not because of adultery. While the implemented confession is a confession that occurred because of the judge's ruling in a lawsuit of the origin of a child.

This is related to Article 287 paragraph (2) of the Civil Law which stated that if one of the crimes as mentioned in article 285 to 288, article 294, or article 322 of the Criminal Law, then the crime can be submitted to the court (Prodjodikoro, 1983). The judge can determine that the evil-doer is the rightful father of a child born of evil deeds based on strong evidence.

Children born from adultery are not allowed to be recognized by the person who commits adultery unless there is a dispensation from the President as stipulated in Article 283 jo 273 of the Civil Law. According to this provision, children born from adultery cannot be forced to confess to the man who fertilized them. It is based on the principle of civil law which determines that the law of marriage must respect the provisions and norms that apply in society.

Article 250 of the Civil Law also stated that "Every child born or raised during marriage obtains the husband as his father." This recognition can be done by a deed made by employees of the Population and Civil Registration Department and must be recorded in the child's birth certificate (Supramono, 1998). Recognition is important because it causes legal consequences in the form of status for children outside marriage and civil relations between the child and the father or the mother who recognizes it as stipulated in Article 280 of the Civil Law which stated: "With the recognition made against a natural child, a civil relationship arises between the child, father, or mother".

As a result of the legal recognition of natural children, there are at least two legal consequences. First, the recognized natural child can use his surname. Second, the child has the right to bequeath either from the father or the mother who admits it. However, under certain circumstances, the confession does not cause a complete result.

It depends on when the confession is made and who admits it. If the confession is done outside of marriage then the confession has a complete effect. However, if the confession is done throughout the marriage of the mother and her husband (not the biological father of the child), then the child outside the recognized marriage is only entitled to use the name of his father (mother's husband/juridical father) alone without the right to inherit from the parent who admits it.

The importance of recognition by the biological father or determination of the origin of the child by the court has legal consequences in the form of legal status, *nasab* (lineage), and other civil relations between the child and his parents, so that between the child and his parents there is a*mahram* relationship, guardians of marriage, mutual inheriting, the obligation of parents to provide a living, finance the education of the child, and others. Similarly, children are obliged to respect and be dutiful to parents and others. In other words, the child has a civil relationship with his parents who authorize him (admit it) especially the civil relationship with the father.

3.2 *The procedure of biological father to admit a natural child that has been recorded in the civil registry based on the marriage certificate*

Recognition of biological fathers to children in birth registration is reviewed from various sides is very important to do. From a legal point of view, recognition in the recording of birth implies the future of the child as well as the identity of the child. A child must know its origin by the determination of the law. However, this recognition is also vulnerable to cause problems, if in the future the biological father of the child wants to admit the child.

The problem is related to the birth certificate that the child already has. When the confession is carried out throughout the marriage of the mother and her husband, as a result the child's birth certificate states that the father is the husband of his mother (juridical father). This happens because the basis of making a birth certificate of them is a marriage certificate either from the Office of Religious Affairs or Civil Registry Agency. On the basis of this issue, it is important to conduct a study related to the procedure or legal steps that can be taken by the biological father when the child has a birth certificate stating that he is the son of his mother's husband (juridical father).

In Article 281 of the Civil Law, it is stated that the way to recognize a natural child, namely in the birth certificate of the child concerned and in the marriage certificate there is. In addition to the way of confession, there are other ways to recognize natural children. In Article 281 Paragraph (2) of the Civil Law, it is stated that the recognition can be made by a certificate made by employees of the Population and Civil Registration Department and must be recorded in the current birth register and subsequently the recognition is recorded in the minutes of the birth certificate of the child concerned. This recognition is made to a child who has been recorded as an out-of-wedlock child in the birth register at the Civil Registry Agency (Subekti, 1995). While the minutes of the certificate is the original deed in the bundle of deeds in the Civil Registry Agency, which is approved by the reporting, witnesses, and officials of the Civil Registry Agency.

Article 55 of the Civil Law confirms that the origin of a child can only be proven by the existence of an authentic birth certificate that has been issued by the authorized official (Abdul, 1997). If the deed does not exist, then the court will hold a closure on the child's origin after a thorough examination based on the evidence that has been qualified. Based on the provisions of the court, the institution of the birth registrar in the legal area of the court concerned issued a birth certificate for the child concerned.

The case of determining the origin of the child including the volunteer case, therefore, in the examination of the case of determining the origin of the child by confession, which must be proven is the terms of recognition as described above, not the correctness of the confession (Syarifuddin, 2006). According to article 287 of the Civil Law, investigating the question of who is the father of a child is prohibited. However, in article 288 of the Civil Law it is stated that investigating the question of who is the mother of a natural child is allowed and, in this case, the out-of-wedlock child must be proven that he is the child born by a mother he mentioned.

According to Article 272 of the Civil Law, every out-of-wedlock child except those born of adultery or desecration of blood, if the father and mother carry out the marriage, then the child becomes a legal child if the father and mother before carrying out the marriage admit it according to the provisions of the Law or the recognition are carried out in a separate deed. In article 280 of the Civil Law, it is stated that with the recognition of a child outside of marriage by the father and mother, the status of the child is valid as other legitimate children (Ustman, 1969).

But this recognition is something different in nature than legitimation. By recognition of a child, it does not become a legitimate child. The natural child, then becomes a legal child, if both parents then marry, after they recognize the child, or if the confession is made in the marriage certificate itself as stipulated in Article 272 of Civil Law (Satrio, 1991; Subekti, 1995). The certainty of a child is really the son of his father is certainly difficult to obtain. In this connection, the law determines and stipulates a grace period of 300 days and the shortest period of 180 days.

To obtain the status of the relationship between father, mother, and child born out of wedlock, the child must be recognized by the father and mother. The confession must be done by authentic deed, expressly, and should not be done in a way that is concluded only. According to Article 224 Paragraph (1) of the Civil Law, the recognition of children outside marriage is null and void if done in some circumstances. First, it is recognized by the man that it is forbidden to marry the mother of the child. Second, a husband who has been married more than 306 days before the birth of a child. Third, a man who is not yet 18 years old, unless the confession occurs at the wedding ceremony. Fourth, without prior consent from the child's mother during her. Fifth, without the prior written consent of the child if he is an adult.

The procedure to change birth certificate after the issuance of the Constitutional Court Decision Number 46/PUUVIII/2010 is as follows. First, the biological father must sue the parents of both the father and the mother who have recognized the natural child as his biological child and have had the birth certificate. Biological fathers must be able to prove it by DNA testing, witnesses who strengthen it, as well as other evidence that strengthens including electronic evidence if any. Second, there are two possibilities that the lawsuit is granted or the lawsuit rejected by the judge. If the lawsuit is granted then the verdict is taken to the Population and Civil Registration Department accompanied by a child's birth certificate. The last step, the Population and Civil Registration Department is to change the record by making a marginal note in the birth certificate because the certificate is made and issued only once by the state.

Based on the above problems, researchers recommend that the Population and Civil Registration Department should register a birth certificate not based on the marriage certificate but based on a birth certificate from a midwife or hospital that has her father's name. In addition, so that the child is not disturbed mentally because he has two fathers who are not the same on his birth certificate, then the first birth certificate needs to be declared missing and then issued a new birth certificate with the name of the biological father. The Office of Religious Affairs at the time of the couple's marriage should add a condition or certificate from the public health center or local hospital that the bride-to-be is not pregnant.

4 CONCLUSIONS

The recognition of biological fathers to children in the birth registry is very important because the recognition is very influential in the life of the child today and the future, especially related to the civil relationship between the child and the father. Confession can be done in four ways. First, it is recorded in the child's birth certificate. Second, it is recorded in the marriage certificate of the father and mother if both are married. Third, it is recorded in the nature of the deed made by the civil registry officer which is then recorded in the birth list according to the date of the making of the deed, then recorded in the birth certificate of the child. Fourth, it is recorded in another authentic deed. In this case, each interested person may demand that this confession be recorded on the child's birth certificate. The legal step of recognition of biological fathers to natural children who already have a birth certificate with juridical parents there are three steps. First, file a lawsuit against the juridical parents. Second, prepare evidence in the form of DNA test results and file two witnesses to testify and strengthen the lawsuit. After that, if the lawsuit is accepted, the Population and Civil Registration Department makes changes to the registrar by making a marginal note in the birth certificate on the basis of the verdict and birth certificate submitted by the biological father.

REFERENCES

Abdul, A. 1997. *Ensiklopedi Hukum Islam*. Jakarta: PT. Ichtiar Baru Van Hoeven.

Afandi, A. 1982. *Hukum waris, hukum keluarga, hukum pembuktian*. Jakarta: Rineka Cipta,

Arifin, M. 2017. Kedudukan anak luar kawin: analisis Putusan Mahkamah Konstitusi Nomor 46/PUUVIII/2010 Tentang Uji Materi Terhadap Pasal 43 Ayat 1 UU Nomor 1 Tahun 1974 Tentang Perkawinan. *Ahkam* 5(1): 117–137.

Darbi, B. A. 2018. Telaah kritis tentang lembaga pengakuan anak (studi kritis antara hukum Islam dan hukum positif). *Al-Bayyinah* 2(1): 91–108.

Hadikusuma, H. 2013. *Metode pembuatan kertas kerja skripsi ilmu hukum*. Bandung: Mandar Maju.

Marzuki, P.M. 2011. *Penelitian hukum*. Jakarta: Kencana Prenada Media Group.

Prodjodikoro, R.W. 1983. *Hukum warisan di Indonesia.* Bandung: Sumur Bandung.

Satrio, J. 1991. *Hukum harta perkawinan*. Bandung: PT Citra Aditya Bakti.

Sirajuddin, S. 2010. Sudut pandang pengadilan terhadap putusan kuasa asuh anak. *Mimbar Hukum dan Peradilan*. Jakarta: Pusat Pengembangan Hukum Islam dan Masyarakat Madani (PPHIMM).

Situmorang, V.M. & Sitanggang, C. 1991 *Aspek hukum akta catatan sipil di Indonesia*. Jakarta: Sinar Grafika.

Soekanto, S. & Mamudji, S. 2004. *Penelitian hukum normatif.* Jakarta: PT. Raja Grafindo Persada.

Subekti. 1995. *Pokok-pokok hukum perdata.* Jakarta: PT. Intermasa.

Supramono, G. 1998. *Segi-segi hubungan luar nikah*. Jakarta: Djambatan.

Syarifuddin, A. 2006. *Hukum perkawinan Islam di Indonesia antara fiqih munakahat dan undang-undang perkawinan.* Jakarta: Prenata Media.

Ustman, F. 1969. *Al fiqh al-Islamy wa tathawwur*. Kuwait: Darul Kuwaitiyyah.

Witanto, D.Y. 2012. *Hukum keluarga (hak dan kedudukan anak luar kawin, pasca keluarnya putusan Mahkamah Konstitusi tentang uji materiil UU Perkawinan*). Jakarta: Prestasi Pustaka.

Empowering Civil Society in the Industrial Revolution 4.0 – Sukowiyono et al (Eds)

Legal protection on the rights of school honorary teachers

S. Awaliyah*
Universitas Negeri Malang, Malang, Indonesia

ABSTRACT: The purpose of this article is to discuss legal protection for school honorary teachers in Malang. This study uses a socio-legal approach. Primary data sources are from school principals, teachers, and institution administrators. Moreover, secondary data sources come from laws and regulations relating to teachers' rights. Data are collected through documentation studies, interviews, and observations. Data analysis used domain and taxonomic analysis. Rights of teachers contained in Law Number 14 of 2005 include the right to get welfare and develop professionalism. This study concludes that the right of school honorary teachers in Malang to get welfare is still not good enough. It is due to the fact that they still get a salary below the city minimum wage. On the other hand, the right to develop professionalism or competence has been obtained well by the teachers.

Keywords: school honorary teachers, teacher salary, teacher welfare, teacher rights, teacher obligation

1 INTRODUCTION

The goal of the state is to educate the nation's life as set forth in Paragraph four of the Preamble of the 1945 Constitution of the Republic of Indonesia. Efforts to educate the nation are regulated through articles, including Article 28 C paragraph (1), namely the right of every person to meet their basic needs, get education, and take advantage of science as a means of improving the quality of life and human welfare, not only for others but also for themselves. Citizens as part of the state have the right to obtain education as stipulated in 5 (five) paragraphs contained in Article 31 paragraph (1) of the 1945 Constitution. The state commitment to education implicitly can be seen in the very large allocation of state budget funds, amounting to 20% of the total state budget (Direktorat Penyusunan APBN dan Direktorat Jenderal Anggaran, 2019). The allocation of the funds is used to finance educational needs, especially for elementary education, construction of infrastructure, and salary of teachers.

Teachers are a crucial component in education. Teachers play roles as educators and means of social justice for students (Rust, 2019). Professionally, teachers are expected to have good qualities to help students to have good attitudes, knowledge and skills as a provision for life. The number of teachers in Malang City based on the database in 2020 is 9,679 people (Kemendikbud, 2020). The number of teachers of the state civil servants is 4,069 people (42%) and non-state civil servants is 5,610 people (58%). Non-state civil servants' teachers consist of permanent foundation teachers, provincial temporary teachers, central auxiliary teachers, and school honorary teachers. Permanent foundation teachers receive a salary according to the school's ability and are entitled to receive certification allowances for those who meet the requirements. Provincial temporary teachers and city/district temporary teachers receive salaries with the help of the provincial and city/district budget. Therefore, the salaries they receive are better.

The problem of this study is related to the status of honorary teachers in Malang, which is around 2,219 people. The salary of the teachers is paid based on the ability of the school, and do not get

*Corresponding Author

 DOI 10.1201/9781003180128-6

assistance from the central or local government. Teachers have the same obligation, namely to provide good and high-quality educational services to students. Meanwhile, Law Number 14 of 2005 concerning Teachers and Lecturers regulates various rights of teachers to get a decent salary and quite a lot of other rights. This study is different from the previous study on foundation honorary teachers using the legal basis of Law Number 13 of 2003 concerning Manpower (Rifai, 2019). The discussion is based on Law Number 14 of 2005. This regulation does not exclude or do not divide the status of teachers as Civil Servants, temporary teachers, permanent teachers, auxiliary teachers, honorary teachers, or other teachers. Based on this, this paper will discuss the protection of the right to welfare and professional development of school honorary teachers.

2 METHODS

This study used socio legal methods. It aimed at revealing the behavior of schools or foundations in providing salaries to school honorary teachers in the schools (Tjandra, 2013). The study used a case study approach to capture data on the welfare of each school's honorary teachers (Soekanto, 1986). The research was conducted in Malang City, East Java Province, which is known as the city of education. Primary data were from school honorary teachers, school principals, and foundation administrators. While secondary data came from regulations concerning teachers, teacher learning plan documents, information about school honorary teachers on the internet, books, journals, and dictionaries. The data were collected through interviews, documentation study, and observation. The data analysis used was domain analysis and taxonomic analysis (Bungin, 2015).

3 RESULTS AND DISCUSSION

3.1 *The right to earn income for school honorary teacher's welfare*

The salary earned by honorary teachers is still very low, below the City/Regency Minimum Wage. The average fee earned is between IDR 20,000 and IDR 75,000 per hour, which is calculated based on the number of teaching hours in one month, not per meeting. Take, for example, if the fee is IDR 40,000 per hour, and a teacher is teaching 24 hours per week, then the salary earned in one month is IDR 40,000 × 24 hours = IDR 960,000. This salary is far below the Minimum Work Wage based on the decree of Governor of East Java Number188/568/KPTS/013/2019 concerning Regency/City Minimum Wages in East Java. The minimum wage is remuneration for work for the performance of workers or employees (Wihastuti & Rahmatullah, 2017). The salary of non-certified school honorary teachers is very far below the Minimum Wage of Malang City, which is IDR 2,600,000.

Various salaries received by school honorary teachers are influenced by the tuition fees paid by students and the number of students. School fees include payments paid at the beginning of the school year and monthly fees. Favorite schools with high achievements tend to get more students and higher costs. Another example is Montessori school which relies on the use of learning media (Rohmatu, et al. 2020). The creation of the media needs a large fund. As a result, the school fee is also high-priced. It is more likely that the more expensive the school fee, then the higher salary the teachers get. On the other hand, undeveloped schools have a little interest and the school fees are low. Consequently, the salary of the school honorary teachers is also low. School conditions greatly affect the salary of school honorary teachers (Fauzi & Syafar, 2017). This is in accordance with Article 15 paragraph (3) of Law Number 14 of 2005, which stipulates that teachers' salary in schools run by the community is adjusted to the capabilities of each school.

School honorary teachers also get a transportation allowance. It is between IDR 30,000 to IDR 100,000 per month. The calculation of the transportation allowance depends on the school institution where the teacher teaches and the number of teaching hours. There are two ways for calculating the transportation allowance, namely per meeting and per month. Social security received by school honorary teachers is health insurance. It is in the form of an insurance card that is valid for one year. The school supports the insurance by two ways, in accordance to the ability of the school. First, the school signs up for insurance and pays the insurance dues without cutting the salary of the teachers. Second, the school only registers for insurance and the teacher will pay the insurance dues.

School honorary teachers have the right to carry out other activities outside of teaching hours at schools in which they usually receive a certain amount of honorarium. Those other activities are temporary in nature which are carried out once a semester or once a year outside of teaching hours. Some of those activities include: (a) compiling exam questions, (b) being an examination supervisor, (c) test evaluation, (d) being a homeroom teacher, and (e) meetings at school or outside of school. In compiling questions, supervising exams, and correcting exams, the teacher will get an honorarium of IDR 15,000 to IDR 40,000. Each month, a homeroom teacher gets an honorarium of around IDR 30,000. The amount of honorarium for meetings or workshops in schools cannot be described on average because the activities vary widely.

This condition is far different from the income or salary of civil servant teachers. School honorary teachers only receive a monthly salary or honorarium based on the number of teaching hours; no functional allowances, professional allowances, and others. Meanwhile, the requirements to become teachers are all the same, which is having a bachelor degree according to the field of science and others. This difference is strongly influenced by several things, namely institutions, regulations that provide protection, and the culture of implementing the regulations (Friedman, 2019). Existing schools or institutions have not been able to provide school honorary teachers a salary or honorarium. School honorary teachers have not gotten adequate legal protection in relation to the salary which affects their welfare. Uncertainty about the status of school honorary teachers has resulted in education providers being able to pay very minimal wages for various reasons. This shows the legal culture of Indonesian society, namely social forces that can be decisive in implementing law (Erwin, 2013).

The income of school honorary teachers as a profession viewed from the Manpower Act is still far from standard. Income is closely related to welfare, which is a fulfillment of both material and non-material needs that are given outside or within a work relationship in order to increase productivity either directly or indirectly. Welfare is closely related to high income (Awaliyah, 2017). At the level of reality, the right to earn income above the necessities of life for school honorary teachers has not been fulfilled yet. It is due to the fact that school honorary teachers get a salary less than the minimum wage for cities and regions. This also contradicts Article 1 Number 16 of the Teacher and Lecturer Law, which states that the granting of teacher rights is due to their achievements and profession as educators.

The Considered to the Teacher and Lecturer Law states that the existence of the law is an effort to improve teacher quality in a directed, planned and sustainable manner. However, by not regulating school honorary teachers, particularly regarding the teachers' welfare, the quality of teachers is not optimal. The quality of teachers is not only influenced by internal factors such as abilities, skills, and experience, but is also influenced by external factors such as salaries (wages), leadership, work environment, and facilities and infrastructure (Arifin & Barnawi, 2012). Salaries have a significant effect on teacher performance or quality (Sari, 2016; Hendrics, 2015). Meanwhile, the quality of teachers is one of the factors that influence the quality of education (Hidayat, 2016). The achievement of high-quality education cannot be separated from the fulfillment of adequate facilities and infrastructure, the availability of funds, quality human resources, and the welfare of educators (Nikmah, 2013). Thus, it is clear that the quality of education is greatly influenced by the welfare of teachers.

Furthermore, calculation of teacher workload is different from the workload of workers in general (Mustofa, 2018). The implementation of the workload for workers in general is 40 hours per week. One day is eight hours of work by using the five working day system, or seven hours by using the six working day system. For workers who have carried out the workload for 40 hours of work, they are entitled to receive the minimum wage set by the government. These working hours only apply to certain professions, not including teaching professions. For the teaching profession, the implementation of the workload is specifically regulated in Regulation of the Minister of Education and Culture Number 15 of 2018.

The teachers' workload at the base of administration for one week is 40 hours, namely 37.5 hours for effective work days and 2.5 hours for rest. It consists of planning, implementing, and assessing the learning being carried out. Besides, teachers have to provide guidance and various training to students. Certain teachers also get additional assignments, for example as a headmaster or vice headmasters of the school. They are also responsible for a minimum of 24 hours and a maximum of 40 hours for face-to-face meeting for one week. Working hours and face-to-face meeting hours are different. One hour of work is 60 minutes, while one hour of face-to-face meeting is different for each level of education; 35 minutes for elementary school, 40 minutes for junior high school, and 45 minutes for senior high school. Teachers who have implemented face-to-face learning for 40 hours do not mean that the teacher has carried out the obligation of 40 hours of work in one week because one hour of face-to-face meeting is not the same as 60 minutes. Therefore, even though the teacher has implemented 40 hours of face-to-face meeting, he must still be present at the base administrative unit to meet the total working hours, namely 40 hours.

Based on the calculations using a minimum teaching load of 24 hours face-to-face meeting, the income earned between civil servant teachers and school honorary teachers is far different. Based on Government Regulation Number15 of 2019, a civil servant teacher at the lowest class of IIIA with a work period of 0 years gets a base salary of IDR 2,579,400, plus allowances and teacher certification one time of the basic salary. Minimum salary, allowances, and teacher certification are around IDR 6,000,000. While the obligations of civil servant and school honorary teachers that must be carried out are the same, namely (a) developing/making learning tools, (b) implementing learning, (c) evaluating student learning outcomes, and (d) carrying out various activities that support the provision of education in schools (Nurmal, 2018).

The obligation of teachers to prepare learning tools is one of absolute tasks (Novauli, 2015). School honorary teachers are also required to make learning tools that include annual programs, semester programs, syllabus, lesson plans, teaching materials, and evaluation. Teachers in carrying out learning must be creative and innovative in accordance with the character of students. The learning media used must be able to attract students. Therefore, the students have a higher enthusiasm for participating in learning and improving their achievement. Schools that have good performance will become favorite schools (Raharjo & Yuliana, 2016) and get more attention from the community. Therefore, they will send their children to those schools. Schools usually charge the cost of creating media to teachers. Moreover, it often requires substantial funds. On the other hand, school honorary teachers' salaries are low or even very low. Then there will be no use of attractive media, because there is no money to create it.

3.2 *Protection and professionalism development of school honorary teachers*

Teachers' rights based on Law Number 14 of 2005 in Article 14, apart from getting a decent income, are to obtain various rights for development of personal, professionalism, competence, and protection for performance in carrying out their duties. In detail, these rights are the rights to get promotion and appreciation for their achievements, get protection in carrying out tasks, determine passing grade for students, get opportunities to improve competence, utilize infrastructure and facilities for carrying out tasks, get freedom to provide assessments, provide rewards and punishments to students, receive security and safety protection during on duty, provide advice or input related to educational policies, develop academic qualifications, participate in professional organizations, and attend training.

The granting of school honorary teachers' rights related to the development and protection of their professionalism as stated in the Teacher and Lecturer Law is effective. This can be seen in the implementation of the rights and obligations of the teacher. Teachers have the freedom to design learning according to their abilities and desires, carry out assessments, give students rewards and punishments as a coaching effort, participate in competitions, participate in various training/workshops/seminars, continue their studies, and become members of the profession. These various rights can be obtained by school honorary teachers according to the talents, interests and wishes of the teachers.

Protection on teachers as stipulated in Article 39 paragraph (4) of Law Number 14 Year 2005 aims at preventing teachers from getting arbitrary treatments. Take, for example, the provision of unreasonable wages or wages, termination of employment without a clear basis, harassment of the profession as a teacher, the expression of restricted views, and other restrictions that are feared becoming obstacles for teachers in carrying out their duties.

School honorary teachers get all protections sufficiently, except the protection of the salary. The teachers have the freedom to express their various opinions either verbally or in writing, of course they still have to pay attention to the applicable regulations. The school and the Education Office provide protections to school honorary teachers. Take, for example when there are problems between teachers and students or with other teachers, the school or the Education Office always provides assistance or acts as a mediator in solving various cases experienced by teachers.

The dismissal of school honorary teachers can be done with honor and disrespectfully. Teachers can be honorably discharged due to retirement, death, serious incurable illness, and at their own request. Dismissal of teachers with disrespect is because the teacher violates the promise and oath of office, violates the work agreement, and for one month or more successively neglects or does not heed his obligations. Schools cannot terminate work relations without an appropriate reason. Termination of employment for school honorary teachers in Malang City is mostly at the teachers' own request because they work elsewhere or are accepted as a candidate for civil servants.

4 CONCLUSIONS

Law Number 14 of 2005 concerning Teachers and Lecturers regulates teachers' rights related to professionalism and welfare for teachers. The granting of the rights to develop professionalism/competence of school honorary teachers in Malang has been effective. The teachers can participate in various trainings, become members of teacher organizations, participate in various competitions, and others. Meanwhile, related to welfare, it is ineffective yet, because the amount of honorarium the teachers get is still below the city and regional minimum wages.

REFERENCES

Awaliyah, S. 2016. The effectiveness of anti-discrimination laws for job seekers in Indonesia. *Journal of Law, Policy and Globalization* 51(1): 41–48.

Bungin, B. 2015. *Analisis data penelitian kualitatif.* Jakarta: Raja Grafindo.

Direktorat Penyusunan APBN dan Direktorat Jenderal Anggaran. (2019). *Informasi APBN 2019, APBN untuk mendorong investasi dan daya saing melalui pembangunan sumber daya manusia*. Jakarta: Direktorat Jenderal Anggaran.

Erwin, M. 2013. *Filsafat hukum.* Jakarta: Rajawali Pers.

Fauzi, H. & Syafar, D. 2017. Studi tentang kebijakan guru honor sekolah sekolah dasar di Yogyakarta. *Jurnal manajemen Pendidikan Islam* 5(2): 162–172.

Hendricks, M.D. 2015. Towards an optimal teacher salary schedule: Designing base salary to attract and retain effective teachers. *Economics of Education Reviews* 47(C): 143–167.

Hidayat, E. 2014. Faktor-faktor yang mempengaruhi mutu sekolah (Pengaruh dari faktor kinerja mengajar guru dan pemanfaatan sumber belajar). *Jurnal Administrasi Pendidikan* 21(1): 81–88.

Friedman, L.M. 2009. *Sistem hukum (perspektif ilmu sosial)*. (Khozin, M. Trans). Bandung: Nusa Media.

Kemendikbud. 2020. *Guru menurut status kepegawaiannya*. Retrieved from https://referensi.data.kemdikbud.go.id/dashboardgtk/ptk_dash2.php?id=20

Mustofa, M. 2018. Menghitung beban kerja guru. *Jurnal Audi Kajian Teori dan Praktik di Bidang Pendidikan Anak Usia Dini* 3(1): 64–69.

Nikmah. 2013. Pengorganisasian layanan pendidikan boarding school berorientasi mutu. *Educational Management* 2(1): 107–112.

Novauli, M.F. 2015. Kompetensi guru dalam peningkatan prestasi belajar pada SMP Negeri Dalam Kota Banda Aceh. *Jurnal Administrasi Pendidikan* 3(1): 45–67.
Nurmal, L.D. 2018. Perlindungan hukum terhadap tenaga pendidik. *Gorontalo Law Review* 1(1): 67–76
Raharjo, S.B. & Yuliana, L. 2016. Manajemen sekolah untuk mencapai sekolah unggul yang menyenangkan: Studi kasus di SMAN 1 Sleman Yogyakarta. *Jurnal Pendidikan dan Kebudayaan* 1(2): 203–217.
Republik Indonesia. 2005. Undang-Undang Nomor 14 Tahun 2005 tentang Guru dan Dosen.
Rifai, A. 2019. Perlindungan hukum terhadap hak-hak bekerja guru yayasan dengan guru honor sekolah sekolah negeri ditinjau dari Undang-Undang Nomor 13 Tahun 2003 Tentang Ketenagakerjaan. *Jurnal Unizar Law Review* 2(2): 158–168.
Rohmatu, H.P. et al. 2020. Kajian komparatif sistem among dan metode montessori. *Jurnal Ilmiah Pendidikan Pancasila dan Kewarganegaraan* 5(2): 261–270.
Rust, F.O. 2019. Redesign in teacher education: The roles of teacher educators. *European Journal of Teacher Education* 42(4): 523–533.
Sari, W.P. 2016. Pengaruh gaji dan motivasi terhadap kinerja guru pada SMA Swasta Bagan Sinembah, Rokan Hilir, Riau. *Tingkap* 12(1): 65–81.
Soekanto, S. 1986. *Pengantar penelitian hukum*. Jakarta: UI Press.
Tjandra, S. 2013. *Sosio legal bekal pengantar dan substansi pendalaman*. Malang: Intimedia.
Wihastuti, L. & Rahmatullah, H. 2017. Upah minimum provinsi (UMP) dan penyerapan tenaga kerja di Pulau Jawa. *Jurnal Gama Societa* 1(1): 96–102.

Empowering Civil Society in the Industrial Revolution 4.0 – Sukowiyono et al (Eds)

Alternatives to land disputes settlement in Indonesia

L. Sumanto*
Universitas Trisakti, Jakarta, Indonesia

ABSTRACT: This paper aims to provide an overview of alternatives to land disputes settlement in Indonesia. This research uses doctrinal legal research methods, descriptive, and qualitative data analysis. The legal certainty of land rights is an essential requirement for landowners and it will be realized if there is any legal protection of land ownership. The Minister of Agrarian Affairs and Spatial Planning/Head of the National Land Agency has a very important role in preventing land ownership disputes. Alternative settlement of land ownership disputes is regulated by law. There are 2 types of land dispute settlement: by litigation and non-litigation. Recently, the Ministry of Agrarian Affairs and Spatial Planning/National Land Agency will start using electronic certificates, it is expected that land conflicts will be reduced.

Keywords: Indonesia, land disputes settlement, mediation

1 INTRODUCTION

Land that is used by anyone and for any purpose must be based on land rights that are protected by law from any arbitrary (Harsono, 2013). Land disputes in conflict affected settings are often considered as a security threat, to be addressed through mediation and strengthening the rule of law (Tajima, 2013). Land disputes can be caused by several factors to show that intervention of another party led to Indonesia's spike in violence (Mamudji, 2017). Law Number 5 of 1960 concerning the Basic Agrarian Law is a legal basis in land issues and legal certainty in Indonesia (Riyadi, 2017). Based on the statement of the Minister of Agrarian Affairs and Spatial Planning/Head of the National Land Agency, Sofyan Jalil, in 2019 there were 8,959 land dispute cases received by the National Land Agency and he had 3,230 land dispute cases that were successfully resolved, consisting of 760 cases included in the K1 settlement criteria, meaning that the Ministry of Agrarian Affairs and Spatial Planning/National Land Agency issued a cancellation decision letter, a rejection decision letter, or a conciliation decision letter between the disputing parties as a result of the mediation of the case concerned; 888 cases included in the K2 settlement criteria, meaning that the settlement of these cases still requires several other requirements and letter of recommendation; 381 cases included in the K3 settlement criteria, meaning that they still require a letter of instruction, as well as a notification letter that the case is not under the authority of the Ministry of Agrarian Affairs and Spatial Planning/National Land Agency. In 2020, the Ministry of Agrarian Affairs and Spatial Planning/National Land Agency is working to resolve 1,201 land dispute cases. The Regulation of the Minister of Agrarian Affairs and Spatial Planning/National Land Agency Number 11 of 2016 concerning Settlement of Land Cases, where it outlines that land disputes can be resolved by related parties through the Ministry of Agrarian Affairs or Court. In 2021, the Ministry of Agrarian Affairs will start using electronic certificates based on the Regulation of the Minister of Agrarian Affairs and Spatial Planning/Head of National Land Agency Number 1 of 2021 regarding Electronic Certificates. The outcome of electronic land registration is in the form of data, information and

*Corresponding Author

DOI 10.1201/9781003180128-7

electronic documents whereas valid and authentic. The products of this electronic land registration are all stored in the Electronic System Database. With the validity of electronic certificates, it is expected that land conflicts will be reduced.

2 METHODS

This research uses normative doctrinal research methods that focuses on the formulation of laws. Doctrinal research is therefore concerned with the discovery and development of legal doctrines for publication in textbooks or journal articles (Chynoweth, 2008). The method in this research is the statute approach (Hutchinson, 2002). The research specifications are descriptive, namely describing, analyzing, and concluding problems that are the object of research concerning land dispute settlements. Library research has a characteristic that identifies various sources that provide factual information or expert opinions on research questions (George, 2008). The legal material reviewed and analyzed in this research uses secondary data consisting of legislation, books, legal journals, and newspapers (Denzin & Yvonna, 1994).

3 RESULTS AND DISCUSSION

3.1 *Land dispute settlement by the ministry of agrarian affairs and spatial planning/national land agency*

A land registration system with a negative publication system tends to lead to the possibility of land disputes (Ginting, 2016). The Regulation of Minister of Agrarian Affairs and Spatial Planning/Head of the National Land Agency Number 11 of 2016 concerning Settlement of Land Cases is regulated land cases can be differentiated between disputes, conflicts, or land cases. Land disputes are disputes over land between individuals, legal entities, or institutions that do not have a broad impact. Land conflict is a land dispute between individuals, groups, organizations, legal entities, or institutions that have a broad impact. Land cases are land disputes which are handled and resolved through judicial institutions. Every land dispute arises because of several very dominant factors, namely (a) incomplete regulations, (b) non-compliance with regulations, (c) unresponsive officials, (d) incomplete and inaccurate land data, (e) incorrect land data, (f) limited human resources, (g) incorrect land transaction, (h) cheating applicants, (i) overlapping dispute resolution authority from other institutions (Sumardjono & Maria, 2008). Alternative Dispute Resolution is a type of procedure voluntarily used by parties in dispute to resolve their issues (Sullivan & Solomou, 2011). Law Number 30 of 1999 regarding Alternative Dispute Resolution and Arbitration regulates dispute resolution options through deliberation of the disputing parties. The definition of Alternative Dispute Resolution is a settlement of disputes or differences of opinion through a procedure agreed by the parties, a settlement outside court through deliberation, negotiation, mediation, conciliation, or expert judgment (Usman, 2013). The administrative process and the recording of the handling of complaints are distinguished for the two reporting mechanisms namely initiatives from the Ministry of Agrarian Affairs and complaints from the public.

Mediation is a method of resolving disputes and conflicts through a negotiation process with the assistance of a mediator to get an agreement between the parties. Settlement through mediation will be simple and deliberative between the parties concerned and the result will be a win-win solution or to the advantages of both parties. The mediator, as a third party helps related parties find the right solution without coercion. The type of mediator used by the Ministry of Agrarian Affairs and Spatial Planning/National Land Agency is an Authoritative Mediator, being an official who has competence and knowledge of the disputes to be handled (Arwana & Arifin, 2019). Mediator is neutral and independent and cannot be influenced by other parties (Yazici, 2016). The decision-making rests with the parties themselves. The mediator's role includes assisting disputants in defining and clarifying issues, reducing obstacles to communication, exploring solutions and reaching a mutually satisfactory agreement. Conflict resolution is managed by one or several expert advisors or through a mediator.

The findings and complaints were analyzed to find out whether the land cases were under the authority of the ministry. Disputes or conflicts which are within the authority of the Ministry of Agrarian Affairs and Spatial Planning/Land National Agency, related to errors procedures in: (a) the measurement, mapping and area calculation processes, (b) the process of registering or recognizing the rights to former customary land, (c) the process of determining and/or registering land rights, (d) the determination of abandoned land, (e) overlapping rights or certificates of land rights where one of the reasons for the rights is wrong, (f) the process of maintaining land registration data, (g) the process of issuing replacement certificates, (h) to providing information on land data, (i) the licensing process, (j) misuse of space utilization, and (k) other errors in the application of statutory regulations. Other than matters related to the dispute or conflict then the Ministry of Agrarian Affairs and Spatial Planning/Land National Agency is not authorized to handle land cases. However, this Ministry can take the initiative to facilitate a dispute or conflict resolution by mediation. Mediation in this regulation is also carried out for types of disputes or conflicts either within the authority of the ministry or those that are not within his authority. Settlement through mediation can be taken if the parties agree to negotiate on the principle of deliberation to reach a consensus for the benefit of all parties. If one of the parties refuses, the settlement can be resolved through the court instead. Technically, mediation is carried out for a maximum of thirty days and the Ministry of Agrarian Affairs and Spatial Planning/Land National Agency acts as the mediator from the Provincial Land Office or the Regency/Municipal Land Office. If the mediation agreement is made, furthermore, a reconciliation agreement is made based on the Mediation Minutes which is legally binding on the parties.

Furthermore, the reconciliation agreement shall be registered at the Registrar's Office of the local District Court to obtain binding legal force. Mediation is deemed canceled, if after being properly invited three times, the parties or one of the disputing parties is not present. Background of the land case is due to a lawsuit/complaint/objection from individuals or legal entities to certificate of title which stipulated by the officials within the National Land Agency, and it is deemed to be detrimental to their land rights (Sumanto, 2008). Based on the lawsuit, the public wants to immediately obtain an administrative settlement called correction from the authorized official. The head of the National Land Agency has the authority to correct the certificate of title. Once he receives complaints from individuals or legal entities, he will aim to resolve the problem by conducting research and collecting data. Based on the results of the research, it can be concluded temporarily whether the complaint can be processed further or not. If the data submitted to the National Land Agency is still incomplete, the National Land Agency will ask for explanation, data, advice from the Head of Provincial Land Office and the Head of Regency/Municipal Land Office where the disputed land is located. When the data is complete, there will be a re-checking of the proposed matter which includes aspects of procedure, authority and legal application. In order to ensure that the interests of the community are protected by law, if deemed necessary, the Head of Regency/Municipal Land Office can block disputed land. This policy is regulated in Regulation of the Minister of Agrarian Affairs and Spatial Planning/Head of the National Agrarian Agency Number 13 of 2017 concerning Procedures for Blocking and Confiscation.

In practice, an individual or legal entity whose interests have been harmed can submit an application blocking a certificate of land title to the Head of the National Land Agency or through the Head of the Regency/Municipal Land Office to be forwarded to the Head of Provincial Land Office where the land is concerned. The Head of Provincial Land Offices and the Head of Regency/Municipal Land Offices can only determine the status quo or block certificates of land title after there is a stipulation of collateral confiscation from the Court. Land cases which are submitted to the National Land Agency to be resolved, the first alternative of settlement for the disputing parties is through mediation where the National Land Agency acts as a mediator. Where land disputes have reached an agreement, written evidence must be made in a notary deed and minutes of meeting. If after examination it turns out that a certificate of title is correct according to the law and in accordance with the applicable procedures, the Head of National Land Agency can issue a decision to reject claims of a third-party objection to the certificate of title issued. Furthermore, he also can issue a decision rejecting the objection from any concerned third party. As a consequence of the refusal, a certificate of title which had been issued will still be valid. However, notwithstanding to a judge's definite legally binding decision, the Head of Regency/Municipal Land Office through the Head of the Provincial Land Office can propose a decision to cancel a certificate of title to the National Land Agency. The head of the National Land Agency can only cancel a certificate of title if there is a legal error or administrative error in its issuance. The authority to cancel a certificate of title is under the Head of the National Land Agency, including the policies to be taken regarding a judge's decision that cannot be implemented.

Land dispute resolution can also be reached through conciliation. Conciliation is the resolution of conflicts, including land conflicts, mediated by one or more neutral conciliators chosen by agreement of the parties (Priyatna, 2002). The conciliator is obliged to provide advice to the disputing parties to resolve the dispute and he must be registered with the Regency/Municipal Land Office (Winarta, 2013). The conciliator must be able to settle the dispute no later than thirty days after receiving a request for resolution. At the first of the settlement, the conciliator is required to reconcile the parties. If a reconciliation agreement occurs, then a joint agreement is made and then registered in the court where the agreement is made. If one party does not exist to the agreement, the other party may submit an application for execution in the court where the collective agreement was lodged (Haspada, 2019).

3.2 *Land dispute settlement by the court*

In principle, if no agreement is reached through mediation between the disputing parties, as mediated by the National Land Agency, the dispute settlement must be through a court. According to the Regulation of the Supreme Court of the Republic of Indonesia Number 1 of 2016, all civil disputes (including land disputes) which are submitted to the District Court shall seek a settlement through reconciliation with the assistance of a mediator first. The first stage is the pre-mediation process stage: A judge will oblige the parties to mutually mediate problems within 7 (seven) days for parties that are domiciled in the same city/regency as the District Court, or 14 (fourteen) days if one of the parties is domiciled in a different city/regency than the District Court. During mediation process in court, the parties have the right to choose a mediator from the options: (a) judges who are not case examiners, (b) lawyer or legal academics, (c) non-legal profession, (d) judge of examining this case, or (e) a combination thereof, if there are more than one mediator in the mediation process, the mediator's duties are determined and agreed upon by the mediators themselves.

We should choose an appropriate mediator, considering experience, credibility, skills, cost, etc. The second stage is the mediation process stage: The mediator will summon the parties to mediate twice. If after two summons one of the parties or both are absent, then the mediation is deemed to have failed. In the event that mediation is deemed to have failed, the mediator will make a recommendation regarding the failure of mediation to the Panel of Judges handling the case, which will become a consideration for the judge in deciding the case. If the parties are present and actively carry out reconciliation efforts, the mediation process will last no more than forty days. Nevertheless, the mediation period may be extended for a period of fourteen days based on the agreement of the parties. If deemed necessary, mediation can be carried out using communication media. The mediation process must be carried out in good faith. One party may declare withdrawal from the mediation process if the other party is proven to have mediated in bad faith. If the mediation results in a reconciliation agreement, the parties, assisted by the mediator, must formulate an agreement reached and signed by the parties and mediator. Before the parties sign the reconciliation agreement or mediation agreement, the mediator examines the material of the reconciliation agreement to avoid an agreement that violates law, or cannot be enforced or harms any third parties. The parties then return to the judge to notify the reconciliation agreement and propose to the judge that the reconciliation agreement be made in the form of a reconciliation deed which is final and binding. If the settlement through mediation in court between the parties is not reached, the settlement of land disputes must be carried out through court channels.

The next stage is evidentiary hearing by reading the plaintiff's claim and response from the defendant. The plaintiff and defendant will argue each other's answers, and the trial will enter the evidentiary stage. The evidentiary process is time consuming and an arduous journey given that all written evidence, testimony from witnesses and expert witnesses, as well as various other evidences must be presented and examined in court. During the trial process, reconciliation is still a possibility given the arduous process.

4 CONCLUSIONS

Dispute resolution can be carried out through two processes: litigation and non-litigation. Litigation is the resolution of legal issues through the courts. Whereas non-litigation means resolving legal matters outside the court. This non-litigation path is known as Alternative Dispute Resolution. A litigation process in court tends to have a longer time frame due to procedural and administrative issues, and also costly for the disputing parties. On the other hand, a non-litigation process outside court where trial process and decisions are not published, the confidentiality of the disputes of the parties is more guaranteed, therefore settlement of dispute is faster, resulting in a win-win solution agreement. The Ministry of Agrarian Affairs and Spatial Planning/National Land Agency will start using electronic certificates, it is expected that land conflicts will be reduced.

REFERENCES

Arwana, Y.C. & Arifin, R. 2019. Mediation on land disputes settlement on promoting human rights fulfillment. *Jambura Law Review* 1(2): 212–236.

Chynoweth. 2008. *Advanced research methods in the built environment.* United Kingdom: Wiley Blackwell.

Denzin, N.K. & Yvonna, S. L. 1994. *Handbook of qualitative research.* London: Thousand Oaks, Sage Publications.

George, M.W. 2008. *The elements of library research: What every student needs to know.* New Jersey: Princeton University.

Ginting, D. 2016. Land registration in information technology perspective a basis of basic agrarian law in indonesia. *International Journal of Scientific & Technology Research* 5(5): 184–186.

Harsono, B. 2013. *Indonesian land law.* Jakarta: Trisakti University Publisher.

Haspada, D. 2019. Arbitration as an alternative dispute resolution for land disputes based on Indonesian regulation. *International Journal of Innovation, Creativity and Change* 8(5): 294–305.

Hutchinson, T. 2002. *Researching and writing in law.* New South Wales: Published Lawbook Co.

Mamudji, S. 2017. Mediation as an alternative dispute resolution in out of court. *Law and Development Journal* 34(3): 194–209.

Priyatna, A. 2002. *Arbitration & alternative dispute resolution.* Jakarta: Fikahati Aneska.

Riyadi, B.S. 2017. Law of agrarian and resolution effort: A claim of eigendom verponding land. *International Journal of Law* 3(3): 80–88.

Sullivan, E. & Solomou, A. 2011. Alternative dispute resolution in land use disputes – two continents and two approaches. *The Urban Lawyer* 43(4): 1035–1059.

Sumanto, L. 2008. Mediation: The alternative of land dispute resolution in Indonesia. Asia Pacific Mediation Forum Conference, *International Islamic University Malaysia* 8.

Sumardjono, and Maria, S.W. 2008. *Mediation of land disputes, potential application of alternative dispute resolution (adr) in the land sector.* Jakarta: Kompas Gramedia.

Tajima, Y. 2013. The institutional basis of intercommunal order: Evidence from Indonesia's democratic transition. *American Journal of Political Science, 57*: 104–119.

Usman, R. 2013. *Alternative dispute resolution in out of court.* Bandung: PT Citra Aditya Bakti,

Winarta, F.H. 2013. *Legal dispute resolution.* Jakarta: Sinar Grafika.

Yazici, H. 2016. The art of opera in Turkey within the scope of culture-art policies. *International Journal of Humanities, Arts and Social Sciences* 2(5): 159–164.

Empowering Civil Society in the Industrial Revolution 4.0 – Sukowiyono et al (Eds)

The urgency of technology function against social empathy in the digital age

N.A. Kurniawan*
Universitas Negeri Malang Malang, Indonesia

ABSTRACT: Individuals in the digital age have the opportunity to change their physical and social environments according to their needs. Empathy is a mental activity that can affect the physical and social environment. Based on data in the field, there has been a shift in the role of social functions into technological functions in various life activities. Restoring the function of technology is an urgency to maintain individual social empathy in the digital era. This paper aims to provide an overview of the urgency of the function of technology for social empathy in the digital age. This study uses a descriptive approach with literature study methods. Sources of research data are books, journals, and research articles. The data analysis technique used content analysis. The results showed that the function of technology has a reciprocal contribution to social empathy

Keywords: technology, social empathy, digital age

1 INTRODUCTION

The development of digital technology has initiated the advancement of human civilization in the 21st century. The emergence of cutting-edge technological devices with the support of advanced features encourages technological development to be more advanced than before the 21st century. Several forms of technological progress can be seen from the aspects of function and form. Technology in terms of function can be interpreted as all kinds of uses that are owned by a technology device. For example, health measurement devices, educational media devices, security systems and so on, while technology in terms of form aspects can be categorized into small or large forms. For example, mobile devices with computers, memory card devices with hard drives, and so on. According to research data, technological developments have reached a higher level than before. The increase in users of digital technology devices shows that the need for technology functions has influenced social functions that run in daily activities (Kurniawan, 2020). The routines that run in the 21st century now seem to almost use the function of technology in all sectors of life, from the health, education, trade, and other life-driving sectors that work together to carry out social functions in life.

Reality like this finally encourages everyone to interact via the internet network. The interactions that occur form new patterns of communication that are able to influence the cultural and value systems of the community at large. Value systems and culture that develop in society provide space for the formation of empathy in the middle of social life. Empathy is concerned based on sympathy for others (Setyaningrum, 2019). Empathy will occur when someone pays attention to the person

*Corresponding Author

DOI 10.1201/9781003180128-8

they meet and wants help to solve or face a problem. The attitude of empathy has an important role in maintaining the balance of values and cultures that develop in society. Another function possessed by an empathic attitude is to encourage someone to care and help (Darman & Howick, 2019) and to drive a meaningful social life emotionally in various situations and conditions of changing civilization (Noten et al. 2019). Empathy is an urgent need of society in the digital era. The use of various digital devices is carried out by almost every age level. According to research data shows that 43% of early adolescents have been in the category of internet addiction (Sari et al. 2017). Other data shows that in a time duration of 1-180 minutes 57.50% of people use the internet during the week and 51.10% on weekends (Kang, 2019). The internet as a product of the digital age is the dominating technology adopted in human civilization.

Realizing the conditions described in the paragraph above, researchers are interested in conducting research on the urgency of the function of technology for social empathy in the digital era. Researchers become research instruments, meaning that data collection and data processing are carried out by the researchers themselves in this study. This study aims to provide an overview of. the urgency of the function of technology to social empathy in the digital era. Through the study of this research, it is expected to be able to add insight to readers regarding the relevance of technology's function to social empathy in the digital era.

2 METHODS

The writing method used in this paper is literature study. Literature study is a method that helps writers find answers about the purpose of writing based on reading results. Researchers in library research are expected to be able to search, find, and manage information sources that can be used as research data. The source of reading becomes the data used by the author to produce an answer or a solution. The selection of data sources or reading is determined by the criteria for publication of the last 10 years and not more than 10 years and above, except for primary or master reading sources. The steps taken for the literature study are presented at table 1.

Table 1. Literature review steps

Steps	Description
Topic Determination	The author chooses a writing focus that can be obtained based on observations, writing objectives and the author's personal experience.
Collecting Data	The author searches data sources to obtain data that supports the purpose of writing. Sources of data can be found online, namely by accessing relevant scientific sites, and offline, namely by collecting books or reading material that have contributed to writing. The data source is then classified and analyzed according to the required part of writing.
Data Review	The data that has been collected and analyzed is then reviewed to produce information that supports writing. The data review follows the format of the purpose, function and usefulness of the writing that will be produced. This data review will later be outlined in the writing.
Writing a Data Review	The data from the study results are then written down in standard into the intended writing format.

3 RESULTS AND DISCUSSION

Social empathy is a mental response that is oriented towards focusing attention to feel the same situation and condition as the target of empathy in the social environment. empathy itself can be interpreted as an important basic attitude that a person has in carrying out life activities as a social being (Tristy, 2017). Social life that continues to change along with the development of information technology is finally enough to provide challenges and opportunities for someone to maintain an attitude of social empathy in order to remain within the norms of values and norms that apply in a cultural system in the social society of the digital era.

The digital era offers many conveniences through sophisticated applications and information technology tools for the public. The digital era is marked by the increasing use of technological devices in various life activities. The use of technology is carried out by the community, such as communicating through social media, buying and selling transactions with the e-money application, internet-assisted learning and learning processes and many other activities that have implemented the function of technology in social life. The function of such technology needs special attention so that it provides greater benefits than the disadvantages and negative impacts that may arise.

The digital age brings many influences in every aspect of life. Empathy is a part of mental activity that takes place during life in various situations. Empathy in technological life in the digital era is important to be a concern for society so that the function of technology does not reduce the role of social functions that form mental activity, one of which is empathy. Some forms of domination of the function of technology in the social function space are dominance in routine interactions with people in the social environment (Mancil et al. 2016) harmonious relationships among communities (Luo et al, 2016), and relationships of mutual help and need among individuals in the social environment (Gellerstedt et al. 2020). The relationship between empathy and technology can be seen in the following table.

Table 2. The relationship between social empathy and technology in the digital age

Social Empathy	Technology
Social empathy drives one's behavior in carrying out an activity	Technology facilitates a person's behavior to drive physical and psychological activities.
Social empathy is a form of personal response to events that occur in the social environment.	Technology provides information about an event or situation. Its function is to provide alternative choices so that someone is more focused according to data information support.
Social Empathy	Technology

According to research data, 97.2% of students receive a significant direct impact on the empathy attitude of using technology in the digital era (Bukhori, 2020). The results of this study illustrate that individuals with higher educational backgrounds also have a risk of being affected by the function of technology on their social functions. worse conditions will certainly be experienced by individuals with lower educational backgrounds. The impact of such an uncontrolled function of technology can actually be anticipated through various strategic efforts. The goal is that social functions can exist and be able to run side by side with technological functions.

Subsequent research shows that 71.8% of people experience the influence of the use of technology on empathy in their social life (Putri, 2019). The effect experienced by society is certainly different from that experienced by students as in the previous paragraph elaboration. Although they have the same core problem, namely regarding the impact of the function of technology on empathy, society is felt to receive a greater impact in this case.

The influence of the function of technology in people's lives can lead to meaningful conflicts. Meaningful conflict can be interpreted as a condition of inner conflict or occurs within the individual and produces a response in the form of an attitude to oppose reality with the mental dimension that is being felt. Communities with various cultural backgrounds and values certainly have the potential to receive influence due to the dominance of the function of technology, so that cooperation between elements of society is needed in maintaining social functions in the function of technology so that it runs dynamically.

Some forms of the influence of information technology with empathy can be shown in the following table.

Table 3. The influence of technology on social empathy

Social Empathy	Technology
Direct Influence	The immediate effect will generally be for a short time. Several forms of technology's direct influence on social empathy are influencing communication styles, ways of thinking and social emotions.
Indirect Influence	The indirect effect lasts quite a long time because the impact will only be felt in the future. Some forms of indirect effect of technology on social empathy are triggering prosocial appearances in daily activities and triggering physical and verbal aggression.
Direct Influence	The immediate effect will generally be for a short time. Several forms of technology's direct influence on social empathy are influencing communication styles, ways of thinking and social emotions.

Seeing the condition of this terrible influence, efforts to restore the function of technology in maintaining social empathy is important so that people are able to avoid other adverse impacts that can be caused. This effort can be done with various alternative steps that are able to accommodate these conditions. Cooperation of various parties is required in realizing this effort in order to achieve optimal results. One of the ways to collaborate is with parties who have the authority in technology system operations such as communication and information technology and telecommunications provider companies, as well as cooperation between technology users, namely people with various backgrounds of technology needs.

Many alternative forms of effort can be made to restore the function of technology so that it does not dominate social functions, including maintaining social empathy in order to exist are through three business alternatives. The first business step is to return to the original purpose of using the technology itself. As we all know, the purpose of using technology actually serves to facilitate the fulfillment of human needs in a broad and global social environment, including in the digital era. However, this orientation experiences a shift and tends to lead to the fulfillment of personal satisfaction. So that this condition will directly or indirectly affect a person's social empathy attitude. Returning destinations can be initiated by putting instructions for use and reinforcing the rules for how to use in accordance with the regulations. In addition, manufacturers and suppliers of technology facilities can also impose sanctions on potential users of the technology devices they produce so that they can be used according to their functions.

The second business step is to provide education about the actual function of a technology device to users. This business looks simply and will be easy to ignore, because almost all users already know the functions of the technology devices they are currently using. However, to maintain a balance of understanding, it is necessary to periodically educate the users. Forms of educational activities that can be carried out are holding seminars, workshops, talk shows, and activities in the form of mass public delivery. This effort will run optimally if there is cooperation with various supporting parties, so that educational targets are appropriate and effectively accepted. The final alternative as the third step is to emphasize understanding to users about the benefits they will receive when using a technology device wisely and appropriately. One of them is to maintain their social empathy in various situations and conditions of change, including changes in the digital era. In addition to benefits, technology owners or producers can also provide losses that can be caused if potential users use technology for interests that have the potential to affect the functioning of their social functions.

4 CONCLUSIONS

Technological developments in the digital era have an influence on social empathy. The increase in the number of technology users encourages changes in social functions to become technological functions and ultimately affects mental conditions in social life. Empathy is a part of mental activity that appears in the form of attitudes that can be observed by the social environment. Empathy and information technology are two different sides but have a mutually influencing relationship with one another. The empathy that emerges is able to maintain technological developments according to social needs and challenges, while the presence of technology is able to direct social empathy towards progress or vice versa. Based on the research results, it shows that the function of technology actually has a strategic role in shaping life. Easy access to activities in meeting physical and psychological needs, placing technology into an inseparable part of life in every community routine in the digital era. However, the dominance of the function of technology ultimately affects the functioning of the social functions that each individual has in society. One of the social functions that runs in the community is social empathy. This research has finally succeeded in providing an idea that social empathy is important to maintain in digital situations, namely situations where the function of technology is able to become dominant over social functions.

REFERENCES

Ati, N.U. & Abidin, A.Z. 2019. Gadget dan mahasiswa (Studi tentang dampak penggunaan gadget terhadap perilaku mahasiswa di Universitas Brawijaya), *Respon Publik* 13(3): 21–29.

Bukhori, M.R. 2020. *Pengaruh kecanduan gadget (smartphone) terhadap empati mahasiswa UIN Maulana Malik Ibrahim Malang*. Malang: Universitas Islam Negeri Maulana Malik Ibrahim.

Gellerstedt, M. et al. 2020. Do Municipal Facebook Performance and Citizen Satisfaction go Hand in Hand? *Electron: Journal e-Government* 18(1): 30–42.

Hardman, D. & Howick, J. 2019. The friendly relationship between therapeutic empathy and person-centered care. *European Journal Person Centered Healthcare* 7(2): 351–357.

Kang, H.J. 2019. The convergence effect of internet usage levels on oral symptoms of adolescents. *Journal Korea Convergence Society* 10(7): 77–85.

Kurniawan, N.A. 2020. Profesionalitas konselor selama pandemi Covid-19. In *Proceeding Seminar Bimbingan dan Konseling*, 87–91.

Luo, N. et al. 2016. How community interactions contribute to harmonious community relationships and customers' identification in online brand communities. *International Journal Information Management* 36(5): 673–685.

Mancil, G. R. et al. 2016. Effects of iPod TouchTM technology as communication devices on peer social interactions across environments. *ERIC* 51(3): 252–264.

Noten, M.M.P.G. et al. 2019. Indicators of affective empathy, cognitive empathy, and social attention during emotional clips in relation to aggression in 3-year-olds. *Journal of Experience: Experimental Child Psychology* 185(1): 35–50.

Puspita, R.S.D. & Gumelar, G. 2014. Pengaruh empati terhadap perilaku prososial dalam berbagai ulang informasi atau retweet kegiatan sosial di jejaring sosial twitter. *Jurnal Penelitian dan Pengukuran Psikologi* 3(1): 1–7.

Sari, A.P. et al. 2017. Tingkat kecanduan internet pada remaja awal. *Jurnal Penelitian Pendidikan Indonesia* 3(2): 10–117.

Putri, S.M. 2019. *Pengaruh smartphone addiction terhadap empati pada generasi milenial*. Riau: Fakultas Psikologi Universitas Islam Riau.

Setyaningrum, R.D. 2019. *Hubungan antara empati dan kecerdasan emosional dengan perilaku prososial*. (Unpublished Thesis). Malang: Universitas Negeri Malang.

Tristy, W. L. 2017. Penggunaan jejaring sosial dengan konsep heppy (*help people around you*) sebagai strategi BK dalam menumbuhkan empati peserta didik. In *Proceeding Seminar Bimbingan dan Konseling* 1(1): 408–416.

Yeo, L.S. et al. 2011. The role of affective and cognitive empathy in physical, verbal, and indirect aggression of a Singaporean sample of boys. *Journal Psychology* 145(4): 313–330.

Empowering Civil Society in the Industrial Revolution 4.0 – Sukowiyono et al (Eds)

Video games as the digital entertainment platform during the COVID-19 pandemic

A.M. Maheswara* & V. Wibowo
University of Indonesia, Jakarta, Indonesia

ABSTRACT: COVID-19 has had a huge impact on the world community — starting from school children, teachers, or employees. Governments in some countries implement a lockdown system that forces people to work and have activities at home. This was an important momentum for video game developers to contribute more to society. This research focuses on the European video game industry, considering that Europe has a sizable video game industry. This study uses a socio-cultural approach with qualitative methods and literature study. Through this research, it is hoped that it will open up new perspectives and knowledge regarding the contribution of video games in society. Based on the results of this study, shows that video games can be a source of alternative entertainment and educational tools to make it easier for people to carry out their daily activities even though they are at home during the pandemic.

Keywords: COVID-19, pandemic, video game industry, socio-cultural, Europe

1 INTRODUCTION

The existence of video games departs from the postmodernism movement. The entry of the post-modernism era caused many things to experience a shift to be more flexible, non-linear, and diverse. The definition of postmodernism itself is said differently by many postmodernists. According to Daniel Bell (1992), postmodernism is an era of post-industrial society that has switched to information technology systems (Bell, 1992). In other words, postmodernism can be said to be part of the current technological progress, resulting in the emergence of various kinds of innovations in terms of technology and the production of goods. The influence of postmodernism shifted some aspects of the use of technology to works of art. Artworks that were initially known to only involve fine art, painting, music, theater or film, are now more developed with the entry of the video game industry into the category of a work of art. Chris Melissinos as an influential figure in the field of technology said in a Time magazine interview (2015) that the video game industry will prove to be one of the most important art mediums that mankind has ever produced (Melissinos, 2015).

The main objective of this research is to find out how the video game industry in Europe continues to run during the COVID-19 pandemic and how they provide alternative entertainment for the wider community, namely through a video game. With this research, it is hoped that it can provide new insights, knowledge, and perspectives for the community on the video game industry, considering that there are still many people who only see the negative impact of a video game.

2 METHODS

This study uses qualitative research methods and the approach used in this study is a literature study approach. The process of data collection in the preparation of this research includes: First, the process of data collection is collected through books, scientific papers, or electronic media with

*Corresponding Author

DOI 10.1201/9781003180128-9

discussions that discuss the global and European video game industry. Second, the authors also find out previous research that discusses the video game industry globally, especially in Europe. Third, processing the data that has been collected and concluding it into a simple analysis.

3 RESULTS AND DISCUSSION

Technological advances have made video games available to the public at large and the industry has become one of the fastest-growing segments in the global media and entertainment market. Video games were started in the 1950s in the United States in the context of research to create a new, more modern recreation platform. Then, in the 1970s the development of video games accelerated rapidly among young people until the 1980s (Wolf, 2012). At that time, video games were still considered insufficiently connected to the arts. However, over time, video games not only contained elements of entertainment but also added an artistic element to them. The growing video game industry in the United States and Japan led countries in Europe to form their industries in the 1980s and 1990s. Since then, the video game industry in Europe has grown and is able to compete in the video game industry market with the United States and Japan.

Among music, film, and video game industries, the industry that has benefited the most is the video game industry. The Entertainment Retailers Association (ERA) said the market value of video games rose to 3.864 billion Euros, more than double its value in 2007. This makes the video game market bigger than the combined film and music markets for the first time (Parson, 2019). Moreover, when the COVID-19 pandemic is occurring, the video game industry around the world is taking advantage of this pandemic to bridge society through video games so they can still get entertainment without having to leave their homes.

During a pandemic like this, many school children are forced to learn from their homes. However, many schools use video games as a medium for providing information regarding all activities at school. One of the schools in England, Wisbech School, uses the game Minecraft as a tool that allows students to get important information in their school. This project has been initiated by the Wisbech School art teacher, John Birch, since 2018 through the Minecraft Education Edition. For Birch, in these conditions, the use of Minecraft is very important for students to get to know their school and school staff more closely. Many parents also responded to this project with positive responses, because their children can still come to school and interact with each other through the virtual school formed by Birch (Leishman, 2020). Minecraft itself is a video game created by a British video game developer, 4J Studios, and already has more than 400 million players from around the world.

Apart from being a medium for school information, Minecraft is also used by academics in several countries to hold graduations virtually. It is recorded that there are more than twenty universities around the world that have conducted virtual graduation through Minecraft, such as Stanford University, Singapore University of Technology and Design, to MIT (Minecraft, 2020). Recently, one of the universities in Indonesia, such as the Sepuluh Nopember Institute of Technology (ITS), used Minecraft to hold a graduation ceremony at the university. For teachers, students, and families, the Minecraft world is a fun medium to take a real Minecraft Ian (Minecraft resident) graduation.

Although they have been popular since before the pandemic, now video games are becoming more and more popular. NPD Group noted that until 2020, there was a 23% increase in sales of video game consoles in the United States, while sales of iPhone mobile games in Europe in July 2020 increased by 20%. The Nielsen survey said that up to June 2020, 41% of game players in France admitted to playing games more often due to the pandemic (Quartz, 2020). This increase in the number of sales certainly confirms that more and more people are spending their time during the pandemic playing video games. Here, people turn to video games as a means to entertain themselves, as well as to pass the time during a pandemic, especially during a lockdown. Health experts have even allowed playing video games as a stress reliever during a pandemic.

At times like this, video games help people to solve their social needs problems. It is also an aid to improving mental health by connecting people and communities, thereby saving them from anxiety and depression. In one study it was found that the regimen prescribed for playing video games significantly reduced symptoms of clinical depression as measured by the Patient Health Questionnaire 9 (PHQ-9) (Russuniello, 2013). It was also found in other studies that research participants who played video games showed improvements in mood and liver function (Anguera, et al, 2017). Socialization through playing games is a good thing for now. Game developers are now focused on increasing engagement between users, which will also increase in-game micro-transactions. Changing consumer perceptions of video games which are now valued as a form of learning activity and a useful way of spending leisure time is a different view, seeing video games as artistic expressions and a form of modern entertainment, which in turn will make the market of this field continue to grow. From this arose a new trend called "Gamification of Education" because of the educational value offered by video games, where games are used as a valuable and high-tech teaching tool to convey 21st-century skills such as creative thinking, problem-solving, and collaboration (Globe, 2020).

It can be seen that video games are contributing to society during the COVID-19 pandemic and creating good impacts, including improving mental health or reducing stress, increasing innovation in the video game industry, and assisting in the implementation of teaching and learning activities. With the limited number of people to move and do something, the number of events that have been canceled or postponed until an undetermined time, the closure of entertainment centers in various places, and the lack of entertainment that can be obtained, video games have become new digital media for the public to move in their mind, socializing, teaching and learning, expression and imagination

4 CONCLUSIONS

Video games were originally considered entertainment. However, the rapid development of technology has made video games turn into an important thing to bring about change, even solutions to some social problems. Looking at the creative process in the video game industry can make us understand that video games are quite complex in their development. During a pandemic, the video game industry can be said to be one of the industries that have succeeded in becoming digital entertainment for the community. In addition, video games are also used as educational media in several countries. This research can open up new perspectives on the use of video games in everyday life. The author argues that in the future, it is possible that video games can be used as a new social media that can connect life between people around the world.

REFERENCES

Anguera, J.A. et al. 2017. Improving late life depression and cognitive control through the use of therapeutic video game Technology: A proof-of-concept randomized. *Depression and Anxiety* 34(6): 508–517.

Bell, D. 1992. The End of Ideology: On the Exhaustion of Political Ideas in the Fifties. New York: Free Press.

Globe Newswire. 2020. *Global video games industry.* Retrieved from: https://www.globenewswire.com/news-release/2020/09/09/2090627/0/en/Global-Video-Games-Industry.html

Leishman, F. 2020. *Thomas clarkson academy: Wisbech school transformed in minecraft so new students can go on a 'tour'.* Retrieved from: https://www.cambridge-news.co.uk/news/local-news/thomas-clarkson-academy-wisbech-open-19051121.

Melissinos, C. 2015. *Video games are one of the most important art forms in history.* Retrieved: https://time.com/collection-post/4038820/chris-melissinos-are-video-games-art/.

Minecraft Education. 2020. *Students around the world hold graduations in minecraft – and now you can too!* Retrieved: https://education.minecraft.net/blog/students-around-the-world-hold-graduations-in-minecraft-and-now-you-can-too.

Parsons, J. 2019. Video Games Are Now Bigger Than Music and Movies Combined. Retrieved: https://metro.co.uk/2019/01/03/video-games-now-popular-music-movies-combined-8304980/

Quartz. (2020). *Everyone is playing video games during the pandemic.* 2020. Retrieved: https://qz.com/1904276/everyone-is-playing-video-games-during-the-pandemic/.

Wolf, M.J. P. (Ed). 2012. Encyclopedia of Video Games: The Culture, Technology, and Art of Gaming. California: Greenwood.

Empowering Civil Society in the Industrial Revolution 4.0 – Sukowiyono et al (Eds)

Strengthening religious character education through Qiroah extracurricular activities

U. Najah* & R. Umar
Universitas Negeri Malang, Malang, Indonesia

D.N. Wijaya
University of Porto, Porto, Portugal

ABSTRACT: This study aimed to describe the character education strengthening program through *qiroah* extracurricular activities to realize religious values and identify obstacles faced and efforts to overcome them. This was a descriptive qualitative research. Observation, interview, and review of documents were procedures carried out for this study. Miles and Huberman's model of data analysis was used to validate and triangulate data. The results showed that the *qiroah* extracurricular activity program enables strengthening the religious character education through training activities or practice reading the Koran, praying the *dhuha*, *istighosah*, reading *yasin*, chanting *tahlil* and making pilgrimage to the grave. There were several obstacles to implement these activities. A few students came late and some of them did not have enough skills to be involved in those activities. Even so, *qiroah* extracurricular activities must be followed by seventh-grade and eight-grade students. The obstacles to implementing the *qiroah* extracurricular activity program in strengthening religious character education could be resolved properly.

Keywords: strengthening character education, extracurricular, Qiroah, Koran

1 INTRODUCTION

Character education needs to be taught to students in accordance with Law Number 20 of 2003 on the National Education System. The law mandated that national education serves to develop the ability and forms the character and civilization of a nation of dignity in order to educate the life of the nation, aiming to develop the potential for learners to become human beings that believe, have faith, noble character, health, knowledge, creativity, independence, and become democratic citizens and have responsibility. Until the National Education System Law Number 20 of 2003 replaced it, the National Education Law of 1946, which took effect in 1947, did not emphasize character education.

Bad character traits occur occasionally and the most serious one is students involved in a fight against other students. The above social phenomenon needs to be overcome by strengthening character education, this is due to the fact that religious character is eroded by globalization. Globalization has materialized the life of students. They engage more with their mobile phones, social media, and interesting games, instead of practicing their religion such as reading Koran. Contrary to the materialization that can cause secularism, religious practices can lead students to know themselves, others and develop good characters.

Data from the Central Statistics Agency of Jombang in 2019 showed that most of Jombang's residents believe in Islam, the number reaches 97.35% of the population. Therefore, the city is

*Corresponding Author

DOI 10.1201/9781003180128-10

also known as the *santri* city. *Santri* are students who dedicated their time to learning the religious practice of Islam. Jombang Regency has the vision to jointly realize a characterful and competitive of the regency. The vision has been transformed into the following missions: (a) to realize clean and professional governance, (b) to realize a quality, religious, and cultured community, (c) improve the competitiveness of the regional economy based on the potential for local and industrial excellence. One of the missions related to this study is to achieve the religious Jombang community. This mission can be understood as the goal of this community is to improve the Jombang community's quality of best religious character.

Students in *Madrasah* (a model of Islamic School) have diverse characteristics and come from different family backgrounds. They also have different habits. This fact is in accordance with the findings of research by Riza (2019). This research found that in achieving this goal *Madrasah* transfers religious values of learning and habituation of religious behavior. *Madrasah Tsanawiyah* Miftahul Ulum in Jombang also has extracurricular activities such as scouting, *al-banjari*, computers, drum bands, and *qiro'ah*. The choice of these activities proved that *Madrasah Tsanawiyah* Miftahul Ulum (MTs Miftahul Ulum) is an Islamic religious foundation that highlighted the religious character as the main foundation for education. MTs Miftahul Ulum has compulsory extracurricular attended by students in seventh-grade and eighth-grade. Extracurricular it is qiroah, a unique activity in the implementation of extracurricular qiroah in addition to training activities or practice of reading the Koran and learning the rhythm of the art of reading the Koran there are activities *dhuha* prayer, reading *istighosah*, reading the *yasin* and chanting *tahlil*, and making pilgrimage. The compulsory extracurriculars are in order to strengthen the religious characters of the learners.

2 METHODS

This research was a descriptive qualitative study. The aim of this study was to produce descriptive data in the form of written or oral words collected from people and the observable behavior of the participants. Qualitative research was more intended to understand existing phenomena such as a journalism, then develop in the form of writing using various scientific methods. Data sources there were informants, events that were by observing events during *qiroah*, and documents stored in the school, such as program books activities and photos of *qiroh* extracurricular implementation. Data collection procedure could be derived from interviews with the informant, namely the principal, *qiroah* extracurricular coaches, and students who follow *qiroah*; observation of the qiroah extracurricular activities, and photos taken during the *qiroah* extracurricular. Data analysis techniques using Miles and Huberman's qualitative data analysis models. According to Miles and Huberman, as quoted in Sugiono (2016), qualitative data analysis could be done starting from data reduction, data presentation, data conclusion, and data verification.

3 RESULTS AND DISCUSSION

3.1 *Qiroah extracurricular activity program in strengthening religious character education*

There are four programs of *qiroah* extracurricular activities applied in MTs Miftahul Ulum in addition to the main program of practice reading the Koran and learning the rhythm of the art of reading the Koran. Those programs are praying *dhuha*, reading *istigosah*, reading *yasin,* chanting *tahlil* and the grave pilgrimage. According to Awaliyah et al. (2018), character development is one of them with the habit of reading and worshiping in the congregation. Strengthened by the results of research by Hambali and Yulianti (2018) religious extracurricular programs include congregational prayer, the art of reading the Koran, *takhfidzul* Koran. This activity, madrasah, has a purpose, whch is to strengthen the religious character in students because religious values are very important for the provision of students when they have graduated madrasah and still bring a good attitude in accordance with the guidance of the Holy Koran as taught in Islam, as has been applied in the *qiroah* extracurricular program to strengthen the education of students' religious character. In accordance with Megawangi (2004), character education is an effort to educate children to make wise decisions and practice them in their daily lives, so that they can make a positive contribution to their environment.

3.2 *Implementation of qiroah extracurricular activities in strengthening religious character education*

Strengthening religious character education through *qiroah* extracurricular activities in the implementation of religious values manifested in *qiroah* extracurricular activities is very important. The extracurricular activities include exemplary values, disciplined values of worship, Islamic brotherhood values, and the value of sincerity. The implementation of those values aims to strengthen religious character education, so religious values need to be applied in its implementation. These values need to be maintained during the strengthening religious education so that students are accustomed to it and can practice it in everyday life and turn it into habit in its implementation. This research is in accordance with the findings of previous research entitled "Strengthening Religious-Based Character Education" by Suryanti and Widayanti (2018). The study mentioned that the religious-based character education strengthening program can improve the quality of schools starting by carrying out habituation activities in its implementation. religious located in *Laskar Pembela Islam* Malang City through the *Maqoman Mahampang* program and the everyday with Koran program.

3.3 *Constraints faced in implementing qiroah extracurricular activities in strengthening religious character education*

The obstacles faced in implementing *qiroah* extracurricular activities in strengthening religious character education are found in students and schools. Among them there are some students in reading the Koran with the rhythm of art or called *qiroah*. Some students do not have the same skills as other students, so some students become less interested, bored, and distracted, therefore, to strengthen religious characters such as those in implementation, namely excellence, discipline in worship, Islamic brotherhood and sincerity are not maximal because they make the atmosphere less conducive and in order. There are limited places in school so that the *qiroah* practice is divided into two sessions. In accordance with Kesuma (2013), strengthening and developing the values of life that are considered important and necessary so that it becomes a personality that is unique to students, as the values developed certainly have obstacles in their implementation, both from students and schools.

3.4 *Efforts to overcome obstacles in the implementation of qiroah extracurricular activities in strengthening religious character education*

In accordance with the obstacles faced in implementing *qiroah* extracurricular activities at MTs Miftahul Ulum, *madrasah* efforts to overcome obstacles to implementing *qiroah* extracurricular activities in strengthening religious character education are to motivate students to be interested in qiroah extracurricular activities in strengthening religious character education by exemplifying exemplary. By giving *tausiyah* to students through extracurricular activities, students can familiarize themselves with their daily lives. The solution so that students are interested and not bored and busy themselves is to add more than one *qiroah* teacher so that the atmosphere is not crowded and becomes conducive. This research is in accordance with previous research entitled "Strengthening Religious-Based Character Education" by Suryanti and Widayanti (2018) that some efforts can be made to realize these activities are exemplifying exemplary, creating a conducive environment.

The efforts of madrasas in overcoming space limitations by building larger madrasah prayer rooms so that they can be used for *qiroah* training are in accordance with the findings of research by

(Budiyanto, 2018) layers of values and beliefs, as well as layers of assumptions, are manifested in the physical aspect of providing worship facilities and facilities support it, while in the behavioral aspect it is carried out with school programs. This is done to strengthen character education in students so that existing obstacles can be resolved with attitude and skills that must be developed so that students can possess the values of noble character which have become a good culture in madrasah. This research is in accordance with previous research entitled "Strengthening Character Education through Extracurricular Activities in Schools" by Dahliyana (2017). This study has similarities to the study by Dahliyana that shows extracurricular activities with character education, namely as an embodiment of knowledge gained in class with attitudes and skills that must be developed so that students can have in the form of noble character values that have become a culture in the social life of the school.

4 CONCLUSIONS

Based on the results of the research and discussion that has been presented in the previous results and discussions about the Strengthening of Religious Character Education through the *Qiroah* extracurricular. It can be concluded that there are four programs of *qiroah* extracurricular activities in strengthening religious character education at MTs Miftahul Ulum, including training programs or reading practicing the Quran and learning the rhythm of the art of reading the Koran, praying *dhuha*, reading *istighosah*, reading *yasin,* chanting *tahlil,* and making pilgrimage to the grave. The *qiroah* extracurricular program aims to strengthen the religious character education applied to students through extracurricular *qiroah* that can be applied to everyday life, which has interrelated programs, what is taught in the Koran reading program with artistic rhythms will be applied to other programs.

The implementation of *qiroah* extracurricular activities in strengthening religious character education at MTs Miftahul Ulum Ulum covers: (a) practice or practice reading the Koran and learning the rhythm of the art of reading the Koran, including recognizing recitation, knowing punctuation signs, understanding the meaning of verse readings, the length of the reading and learning the rhythm of the art of reading the Koran, (b) *dhuha* prayers are carried out on Saturdays and Mondays at the madrasa yard using mats in the morning before class starts, (c) reading *istigosah, yasin*, and *tahlil* are held in the morning the day before the learning hour starts every Thursday and at the beginning of each month in front of the class, (d) a grave pilgrimage is carried out in Jombang Regency, namely to the graves of the *madrasah* founders and the founding figures of *Nahdlatul Ulama*. This religious values that are manifested in the implementation of strengthening religious education in extracurricular activities including excellence, discipline in worship, Islamic brotherhood, and sincerity.

REFERENCES

Awaliyah, S. et al. 2018. The implementation of Regulation of the Minister of Education and Culture Number 23 Year 2015 Concerning Character Development in the Middle School. *Advances in Social Science, Education and Humanities Research* 251: 367–370.

Badan Pusat Statistik Kabupaten Jombang. 2019. *Penduduk menurut agama yang dianut*. Retrieved from https://jombangkab.bps.go.id/statictable/2020/08/19/5971/penduduk-menurut-agama-yang-dianut-2019.html

Budiyanto, R. 2018. Pendidikan karakter religius berbasis budaya sekolah di SDIT Luqman Al-Hakim Internasional Yogyakarta. *Jurnal Pendidikan Guru Sekolah Dasar* 6(7): 581–592.

Dahliyana, A. 2017. Penguatan pendidikan karakter melalui kegiatan ekstrakurikuler di sekolah. *Jurnal Sosioreligi* 15(1): 54–64.

Gunawan, H. 2012. *Pendidikan karakter*. Bandung: Alfabeta.

Hambali, M. & Yulianti, E. 2018. Ekstrakurikuler keagamaan terhadap pembentukan karakter religius peserta didik di Kota Majapahit. *Jurnal Pedagogik* 5(2): 93–208.

Kesuma, D.D. 2013. Pendidikan karakter. Bandung: PT. Remaja Rosdakarya Offset.

Megawangi, R. 2004. Pendidikan karakter, solusi yang tepat untuk membangun bangsa. Bogor: Indonesia Heritage Foundation.

Riza, A. 2019. *Peran madrasah diniyah dalam pembentukan karakter religius santri: Studi di Madrasah Diniyah Awaliyah Al-Muttaqin Rengging Kabupaten Jepara.* (Unpublished Thesis). Semarang: Fakultas Ilmu Tarbiyah dan Keguruan Universitas Negeri Walisongo.

Sugiono. 2016. *Metode penelitian pendidikan: pendekatan kuantitatif: kualitatif dan R&D.* Bandung: Alfabeta.

Suryanti, E.W. & Widayanti, F.D. 2018. Penguatan pendidikan karakter berbasis religius. In Sabar Setiawidayat et al. (eds.) *Inovasi IPTEKS untuk mendukung pembangunan berkelanjutan; Proceeding Conference on Innovation and Application of Science and Technology, Malang, 12 September 2018.* Malang: Universitas Widyagama Malang.

Republik Indonesia. 2003. Undang-Undang Nomor 20 Tahun 2003 Tentang Sistem Pendidikan Nasional.

Empowering Civil Society in the Industrial Revolution 4.0 – Sukowiyono et al (Eds)

The development of an anti-corruption education module from a contextual perspective in higher education

M. Zakiyah*, P.K. Wulandari & P.K. Dewi
University of Brawijaya, Malang, Indonesia

ABSTRACT: Indonesia has been attempting to stop corruption. This effort is in the level of prevention from a legal perspective. However, the fighting towards corruption should be approached from a social-cultural perspective. This study was conducted at the center of character building. The outcome of this study was to compose a module on anti-corruption education from a contextual perspective. Contextual perspective was used due to (1) meaningful approach, the student could build their knowledge from their experience; (2) direct the student to be the honest and caring generation. The result of the module development is the content of four-chapters of (1) corruption from law, religion, politics, socio-culture, and the ethics of Pancasila; (2) building self-integrity and following the footsteps of the founding fathers; (3) student as the agent of anti-corruption; (4) conclusion. Each chapter are also accompanied by the learning aims, learning materials, and tasks

Keywords: module, anti-corruption education, contextual perspective

1 INTRODUCTION

Corruption is a major problem in Indonesia. In 2020, it was revealed that the biggest corruption case of PT *Asuransi Jiwasraya (Persero)* had caused the state to lose more than 13.7 trillion (Hardiyanto 2020). Ironically, all 40 suspects of the corruption case in 2018 involving members of the Malang City DPRD, have yet to be prosecuted. Besides, the revision of the Corruption Eradication Commission Law (KPK Law) issued by the legislative body (DPR) around October 2019 made the public anxious and led to protests in several cities. Public fear is the potential for the emergence of articles that weaken the KPK, such as the DPR which recommends revoking the investigative and prosecutorial powers that have been the strength of the KPK in eradicating corruption. Yulianto & Zakiyah (2018) explain that the formation of the KPK is still limited to anti-corruption measures and has several weaknesses, such as injustice in the criminal justice of corruption, fines and repayments of state cash that are not proportional to the number of state losses, and the lack of national integrity in fighting corruption. Thus, support from all parties including academics as agents of preventing acts of corruption is needed.

What needs to be appreciated is that awareness of the prevention and eradication of corruption by Indonesia has produced few positive results. According to the 2018 Transparency International (TI) survey among ASEAN countries, Indonesia's Corruption Perception Index (CPI) was in 4th place with 38 points on a 0–100 scale, 1 point higher than the previous year and was able to shift Thailand's position. Even though it has not been able to occupy the top position, it can be a positive signal in efforts to prevent and eradicate corruption in a sustainable manner (Transparency International in BPS 2018).

Furthermore, Wijoyanto in Sofia (2011) revealed several approaches to action against corruption, namely: a lawyer approach, a business approach, a market/economic approach, and a socio-cultural

*Corresponding Author

 DOI 10.1201/9781003180128-11

approach. As academics, we have a responsibility to participate in fighting corruption by pursuing a socio-cultural approach.

Kadir (2013) argues that anti-corruption learning still focuses on a legal approach. Thus, learning innovations that are more interesting and able to raise students' awareness and encourage them to have an attitude of anti-corruption are needed. Contextual Teaching and Learning (CTL) is considered appropriate to be one of the innovations in anti-corruption education because it provides opportunities for students to seek their own experiences in the field. Students not only learn things that are cognitive, but this learning is directed to touch the affective realm so that they are able to build student character.

In implementing CTL, researchers initiated the development of an anti-corruption module to make it easier for researchers to apply this approach on students. The development of an anti-corruption module with a contextual perspective, in addition to containing material on corruption, is complemented by instructions on how students seek experience in the field regarding actions that are allegedly leading to corruption. The development of the anti-corruption module will be integrated in the General Compulsory Subject (MKWU), especially Citizenship and Indonesian Language Education at Brawijaya University. The development of the anti-corruption module with contextual insight is expected to be able to contribute to building the character of anti-corruption integrity to UB students.

2 METHODS

This study uses the Borgh & Gall development model (1983 in Silalahi 2017) with four core research stages, namely (1) the pre-development stage, (2) the development stage, (3) the trial stage, and (4) the revision stage. The research, which is located at Brawijaya University, began with a pre-development stage consisting of a preliminary analysis and analysis of student characteristics. The development stage was modeling. The trial phase was small group trials for Universitas Brawijaya students who were taking Indonesian Language or Citizenship Education courses. The revision stage is model revision activities and the preparation of the final product (Table 1).

Table 1. Score achievement actions.

Value	Actions
85–100	Implementation, meaning that there is no need for revision. Revisions are only in the form of minor revisions as suggested by the validator.
70–84	Implementation with a few revisions based on notes of test subjects.
55–69	The revisions are based largely on the notes of test subjects.
<55	Big revision. Researchers can replace or change the aspects presented in teaching materials.

3 RESULTS AND DISCUSSION

3.1 *Description of the contextually informed anti-corruption module*

As stated in the introduction, this research aims to produce an anti-corruption module based on contextual learning. Zulfa (2019) states that contextual learning is one of strategies used for learning values, especially life values in society. This study focuses on student-oriented learning. The students should find the related concept or theory and compare it to the real world. Therefore, this would affect the student's psychomotor to absorb the good value in their life (Sanjaya 2006; Suyanti 2010 in Sinaga & Silaban 2020). Contextual learning seems to be the right choice to apply anti-corruption education. It is believed that the students would be the dynamo of social integrity.

In order to develop an anti-corruption module based on contextual learning, seven components of contextual learning were used, namely: constructivism, inquiry, questions, learning community, modeling, reflection, and authentic assessment. In order to realize these seven components, this module is presented in four chapters. The first chapter presents "Corruption in the Perspective of Law, Religion, Politics, Socio-Culture and Pancasila Ethics". This chapter discusses the nature of corruption from various perspectives. In this presentation, even though it is cognitive, students will be asked to build their own understanding of the criminal act of corruption (constructivism) through practice questions. The understanding that exists in students so far will be strengthened by the theory in chapter one and the practice questions in this chapter. Furthermore, in chapter II, it discusses "Building Self Integrity and Emulating National Teachers". In this chapter, examples of national teachers and internalization of anti-corruption values are presented as modeling. Students are expected to get models or examples of anti-corruption figures from the nation's teachers. In order to strengthen the impression of modeling, in the training section students are asked to look for millennial models as anti-corruption figures who can inspire them.

Furthermore, chapter III discusses "Students as Anti-Corruption Agents (Inquiry-Based Case Analysis Model)" This chapter embodies the components of inquiry, learning community, questions, and authentic assessment. In this chapter, students are asked to carry out field investigations related to the misuse of service procedures around them. Students are expected to discover by themselves (inquiry) the misuse of service procedures that is the forerunner of corruption or mini corruption. This activity was originally carried out in groups so as to include a learning community component and an authentic assessment because students will be asked to evaluate and assess the results of their own and other groups' fieldwork. This activity will certainly give students experience to ask questions both to themselves, to the team, as well as to the servants (community/campus service providers).

This module has components for each chapter in the form of (1) learning objectives, (2) material, and (3) exercises. The learning objectives of the first chapter are (1) explaining corruption from a legal perspective, (2) explaining anti-corruption from a religious perspective, (3) describing corruption in the circle of law and power, (4) describing the anti-corruption culture and social sanctions, and (5) presenting Pancasila ethics and attitudes of corruption. This objective is consistent with the learning material which contains (1) introduction, (2) corruption from a legal perspective, (3) anti-corruption from a religious perspective, (4) anti-corruption culture and social sanctions, and (5) Pancasila ethics and corruption attitudes. To be able to measure learning achievement, the exercise at the end is arranged in the form of questions that show the reasons for the occurrence of criminal acts of corruption. From here, it will be seen that the building of students' understanding of the criminal act of corruption and the reasons that can be related to the perspective of law, religion, Pancasila ethics and anti-corruption attitudes, to why there is an anti-corruption culture.

To demonstrate students' understanding of the nature of corruption, the first chapter begins with a question on student understanding. This question also serves as the foundation for constructivism, that students build their understanding of corruption on the building of their previous understanding.

Furthermore, Chapter II has learning objectives as follows: explaining the value of anti-corruption and character education in UB, internalizing religious values (Islam), internalizing socio-cultural values, internalizing anti-corruption values in Pancasila, and imitating the nation's teachers. The material in this chapter is in accordance with the learning objectives. The exercises in this chapter are in the form of exercises to relate theories that have been studied with realities in the field, such as that religion teaches anti-corruption values, but there are figures known to be religious who are involved in criminal cases of corruption. Students are also asked to look for cultural norms that have the potential to foster a culture of corruption, cultural norms that can be used as sanctions for criminal acts of corruption and which can be used as preventive measures for corruption, and finally look for millennial figures who are considered capable of becoming anti-corruption figures.

To be able to explore students 'initial understanding, this chapter is presented with opening questions that can explore students' understanding of corruption in cultural, social, and religious norms, internalization of anti-corruption values, and national figures with high integrity. This question ensures that the new understanding constructs built in this chapter build on students' initial understanding of the material taught in this chapter.

Furthermore, the learning objectives in chapter III are describing the concept of contextual learning, connecting contextual learning and anti-corruption education, and conducting reporting and investigations. The material in this chapter discusses and invites students to carry out these learning objectives. The exercises in this chapter are in the form of investigative reports on allegations of abuse of service procedures around students. This exercise is also a form of inquiry, learning community, authentic assessment, and questions in contextual learning. To ensure that their understanding is constructive, this chapter begins with questions about students' knowledge so far about contextual learning, the potential for contextual learning as anti-corruption education, and investigative reporting practices.

The final chapter is Reflections and Closing. This chapter provides the opportunity for students to write and reflect on what they have done. This includes reflections on findings in the field. From this reflection, students are asked to rewrite the learning experiences they have gained through this activity, construct a new understanding of the criminal act of corruption, so that it is hoped that there will be an internalization of anti-corruption values embedded in the minds of students. Exercises at the reflection stage are also exercises used in the pretest and posttest to measure the effectiveness of the module.

3.2 *Refinement of the contextually informed anti-corruption module*

To be able to assess the feasibility of this contextual learning-based anti-corruption module, a validation test was then carried out on two experts, namely the module development expert and the anti-corruption study expert. Experts in developing focus modules assess the contextual aspects of learning, presentation, content, and the language used in this module. Meanwhile, anti-corruption research experts focus on corruption and anti-corruption material, presentation, and the language used.

After that, the module was revised based on the expert test results. The module will also be tested for its effectiveness in small groups online given the pandemic conditions that limit the possibility of face-to-face becoming virtual ones.

The following is the recapitulation of expert validation test scores on the anti-corruption module based on contextual learning which is presented in Table 2.

Table 2. The recapitulation of validation test scores on the anti-corruption module based on contextual learning.

No	Validator	x	%	Note
1.	Anti-corruption studies expert	2.95	73.8%	Valid
2.	Module development experts	3.33	83.4%	Valid
3.	Students	3.43	85.8%	Very Valid
	Average	3.24	81%	Very Valid

Based on expert judgment, this module has been declared valid. The average score of all aspects based on the expert's assessment of anti-corruption studies is 2.95 with a percentage of 73.8%. This means that this teaching material is valid and feasible to be implemented with several revisions according to the direction of the expert. Meanwhile, based on the expert's judgment, the preparation of this teaching material module got an average score of 3.34 with a percentage of 83.4%. This means, as the results of the assessment of the anti-corruption study expert, this module is feasible to be implemented and is valid by making several revisions according to the expert's notes. Based on students' judgment, this module got a score 3.43 or 85.5% in percentage. The total score of the two experts in all aspects is 3.24 with a percentage of 81% which indicates that this module is very valid and feasible to be implemented with several revisions.

Furthermore, based on each aspect, here are the results of the validator's assessment of the contextual learning-based anti-corruption module. In the contextual aspect of learning, the following is a recapitulation of expert judgment which is presented in Table 3.

Table 3. The score of contextual learning's aspect.

No	Validators	x	%	Note
1.	Anti-corruption studies expert	2.85	71.4%	Valid
2.	Module development experts	3.31	82.8 %	Valid
	Average	3.08	77.1%	Valid

Based on the results of the recapitulation of anti-corruption study experts, the contextual learning aspects of this module are rated on average 2.85 with a percentage of 71.4%. This means, according to an anti-corruption study expert, this module is valid and feasible to be implemented with revisions in several parts, such as adding cases close to students. According to module development experts, this module is rated at an average of 3.31 with a percentage of 82.8%. This means that this module is valid and feasible to be implemented with several revisions.

In the presentation aspect, a recapitulation of expert judgment can be seen in Table 4.

Table 4. The score of presentation aspect.

No	Validator	x	%	Note
1.	Anti-corruption studies expert	2.88	72%	Valid
2.	Module development experts	3.68	92%	Very Valid
3.	Students	3.4	85%	Very valid
	Average	3.32	83%	Very Valid

Based on the expert's assessment of anti-corruption studies, this module gets an average score of 3.6 with a percentage of 72%. This means that this module is valid and feasible for implementation with revisions according to the notes, such as the preparation of modules which are judged to begin with abstract (difficult) material to concrete (easy) material. Even so with the module development experts who rated an average of 4.6 and a percentage of 92% for the presentation of this module. This shows that this module is very valid and feasible to be implemented with a limited revision, if necessary. Based on students' judgment, this module got a score 3.4 or 85% in percentage. In the aspect of module content, the following is a recapitulation of the assessment of the anti-corruption study expert and the module development exper can be seen at Table 5.

Table 5. The score of modul content's aspect.

No	Validator	x	%	Note
1.	Anti-corruption studies expert	2.96	74.2%	Valid
2.	Module development experts	3.31	82.8 %	Valid
3.	Students	3.38	84.6%	Very valid
	Average	3.21	80.4%	Valid

Based on the recapitulation, the anti-corruption study experts considered that this module was valid with a value of 2.96 and a percentage of 74.2%, which means that it is feasible to be implemented with several revisions. According to module development experts, this module is in the valid category with a value of 3.31 with a percentage of 82.8%. Based on students' judgment, this module got a score 3.38 or 84.6% in percentage. This means that this module is worth implementing with several revisions.

In the language aspect, a recapitulation of the assessment of the anti-corruption study expert and the module development expert can be seen at Table 6.

Table 6. The score of language aspect.

No	Validator	x	%	Note
1.	Anti-corruption studies expert	3.28	80%	Valid
2.	Module development experts	3	75%	Valid
3.	Students	3.52	88%	Very Valid
	Average	3.24	81%	Very Valid

Based on the results of the recapitulation of the language aspects, it is known that the anti-corruption study experts assess that this module is valid and feasible to be implemented with several revisions with an average value of 3.28 and a percentage of 80%. Meanwhile, according to the module development expert, the language in this module received an average score of 3 with a percentage of 75%. Meanwhile, students' judgement for this module got a score 3.52 or in percentage 88%. This means that this module is very valid and the service is implemented with several revisions to the language aspect.

Based on the test results from the anti-corruption study expert, there are several notes and revisions that must be made, which include typos, incorrect use of terms, and theoretical errors. In fact, invalid laws are no longer called laws. There are also statements that there are laws that are no longer valid and that other laws are removed, even though these laws are still valid. False statements or theoretical justifications are thus eliminated. Furthermore, there is input to add female teacher figures. In this case, revisions were made by adding female figures, such as "Najwa Shihab" who was concerned about guarding suspected corruption cases in Indonesia. Additionally, according to the expert, this module is structured from abstract/difficult to concrete/easy. In fact, the principle of developing modules is to start from the easy ones to the difficult ones. For this, the module composition is rearranged so that it goes from easy to difficult. Finally, according to the expert, this module needs to be added to cases more relevant to students. Therefore, this module will add examples that are close to students.

Based on the validation test on module development experts, there are several notes. First, the use of language errors, such as the misuse of the conjunction "then" at the beginning of a sentence and the use of the pronoun "this" should be "that". For language errors, revisions are made according to the direction of the expert. Second, the systematic error/sequence of the arrangement of chapters, as in chapter I points C and D, which according to experts should be reversed. In this error, revisions are made by following expert input. According to the module development expert, the contextual learning-based anti-corruption module is interesting because it uses a variety of study perspectives to analyze corruption studies

4 CONCLUSIONS

The specification of the contextual learning-based anti-corruption module contains four chapters, namely (1) Corruption in the Perspective of Law, Religion, Politics, Socio-Culture and Pancasila Ethics, (2) Building Self Integrity and Emulating National Teachers, (3) Students as Anti-Corruption Agents (Case-Based Analysis Model Inquiry), and (4) Reflection and Closing. Each chapter is equipped with learning objectives, materials, and exercises. This module uses seven components of contextual learning, namely constructivism, inquiry, questions, learning community, modeling, reflection, and authentic assessment. Constructivism is developed in each chapter in the form of questions related to student understanding before studying the material in each chapter. Inquiry, questioning, learning community, and authentic assessment are developed in Chapter III. Modeling is developed in chapter II. Meanwhile, reflection is developed in chapter IV. Based on product development and validation tests from anti-corruption studies experts and module development experts, the contextual learning-based anti-corruption module is considered valid and feasible to be implemented with several revisions.

REFERENCES

Hadin, A.F. & Pahlevi, R. 2016. Desain bahan ajar pendidikan kewarganegaraan berbasis pendidikan anti korupsi di perguruan tinggi. *Jurnal Moral Kemasyarakatan* 1(2): 162–172.

Hanoum, R.N. 2014. Mengembangkan keterampilan berpikir tingkat tinggi mahasiswa melalui media sosial. *Edutech*, 13(3): 400–408.

Hardiyanto, S. 2020, *Selain jiwasraya, berikut kasus korupsi terbesar di indonesia.*

Kadir, A. 2013. Konsep pembelajaran kontekstual di sekolah. *Jurnal Dinamika Ilmu* 13(1):17–38.

Krathwohl, D.R. 2002. A revision of Bloom's taxonomy: An overview. *Theory into Practice* 41(4): 212–218.

Sanjaya, W. 2006. *Strategi pembelajaran berorientasi standar proses pendidikan.* Jakarta: Kencana Prenada Media Group.

Saripudin, D. & Komalasari, K. 2016. Culture-based contextual social studies learning for development of social and cultural values of junior high school students. *The Social Sciences* 11(23): 5726–5731.

Schunk, D.H. 2012. *Learning theories: An educational perspective.* 6th Edn. New York: Pearson Education.

Silalahi, A. 2017. *Development research (penelitian pengembangan) dan research & development (penelitian & pengembangan) dalam bidang pendidikan/pembelajaran.* (Unpublished Dissertation) on doctoral program Universitas Negeri Medan.

Sinaga, M. & Silaban, S. 2020. Implementasi pembelajaran kontekstual untuk aktivitas dan hasil belajar kimia siswa.*Jurnal Gagasan Pendidikan Indonesia* 1(1): 33–40

Sofia, A.I. 2011. *Model pembelajaran mata kuliah anti-korupsi. dalam pendidikan antikorupsi untuk perguruan tinggi*. Jakarta: Kemdikbud.

Transparansi Internasional. *Corruption perceptions index*. (Online).

Trianto. 2009. *Mendesain model pembelajaran inovatif progresif.* Surabaya:Kencana.

Wicaksono, S.R. 2011. Strategi penerapan domain afektif di lingkup perguruan tinggi. *Jurnal Pendidikan* 2(2) September 2011: 112–119.

Yulianto & Zakiyah, M. 2018. Reaktualisasi Zuhud Al Ghazali sebagai model pendidikan antikorupsi pada pendidikan anak usia dini dan dasar. proceeding on international seminars Al Ghazali, Jakarta, January 18, 2018.

Zulfa. 2019. Pembelajaran kontekstual nilai-nilai karakter mata kuliah budaya Minangkabau sebagai penguatan identitas kebangsaan. *Journal Moral and Civic Education* 3(1): 38–47

Empowering Civil Society in the Industrial Revolution 4.0 – Sukowiyono et al (Eds)

Strategies for internalizing the value of nationalism in tunagrahita students

P. Sekarsari*
Universitas Negeri Malang, Malang, Indonesia

ABSTRACT: This study described strategies, constraints, and efforts for internalizing the value of nationalism to tunagrahita students. Tunagrahita is a term used to refer to children who have below-average intellectual abilities. The research was conducted using a qualitative approach from observation, interviews, and document studies. The results showed that the strategy was carried out through programs, resource utilization, policies and regulations, and optimal learning. The obstacles faced were the lack of support from parents, differences in character in students, and the difficulty of finding teachers that met school standards. Efforts to overcome these obstacles were providing insights to parents, memorizing the character that each student has, and increasing the creativity of teachers.

Keywords: nationalism, disabilities, internalization

1 INTRODUCTION

The value of nationalism needs to be developed in students with no exception and also for people with special needs. A previous study of visually impaired students showed that these students are used to behaving positively so as to reduce the chances of students behaving negatively in society. Based on research conducted by Ardianto (2013), behavior that is considered negative by the community is actually the result of lack of handling and organizing specifically in people with disabilities, so that students tend to behave as they please.

The visually impaired belong to the group of children with special needs. Tunagrahita is a term to refer to children who have below-average intellectual abilities. Deafness according to Abdurrahman quoted by Wantah (2007) is a disability. Children can lose mental development so children with visual impairment have the main characteristic of being weak in thinking or reasoning that results in learning ability and social adaptation being below average. The definition of visual impairment according to Desiningrum quoted by Rahayu (2019) is a child with special needs who is behind in mental, physical, emotional, and social development who needs special treatment in order to develop at the maximum ability.

This study answered three questions. First, how is the strategy of internalizing the value of nationalism in visually impaired students? Second, what are the obstacles faced in internalizing the value of nationalism in visually impaired students? Third, how to overcome the constraints in internalizing the value of nationalism in visually impaired students? Based on those questions, the purposes of this study were to (1) describe the strategy of internalization of the value of nationalism in visually impaired students, (2) describe the constraints faced in the internalization of the value of nationalism in visually impaired students, and (3) describe efforts to overcome constraints in the internalization of the value of nationalism in visually impaired students.

---

*Corresponding Author

DOI 10.1201/9781003180128-12

Strategy is one's internal ability to think, solve problems, and make decisions, taking into account all the resources that an organization has. Strategies in the context of education involve all components, especially teachers and students to be able to carry out the learning process. Strategies according to Wena (2012) means ways and art of using resources to achieve certain goals. A person's character can change after going through several processes, one of which is by internalization. Internalization is the process of entering a good character of a person, in the hope that a person can apply the character that has been included in the internalization process in everyday life. Internalization is a process that takes time and produces new behavior from the process (Wantah, 2007). The internalization process takes time because inserting the character of a person is not easy, considering that a person is also affected by various factors, such as environmental factors both in the community, family, and school. Behaving or having good character is the purpose of the internalization process because in this process one is expected to apply the character that has been included in the internalization process.

2 METHODS

In general, this study aimed to analyze a phenomenon or event that becomes a social activity. This was a qualitative research that according to Bogdan & Taylor was "a research procedure that produces descriptive data in the form of written words from several people and observable behavior" (Moleong, 2016). The use of qualitative method of this study means to get in-depth data and contain meaning. The meaning in question is something behind the phenomenon that is not yet known, or it can also aim to gain insight into a little-known thing. Data collected consist of words, images, and not numbers.

Data were collected using observation, interviews, and document studies. According to Miles and Huberman (Sugiyono, 2015) in a qualitative study, analysis is conducted interactively and takes place continuously until complete and the data is already at saturation point. The interactive data analysis described by Moleong (2016) is based on a paradigm view of positivism. At the time of the interview, an analysis of the informant's statements was carried out. When the statement from the informant after being analyzed was not satisfactory then continue with the question again until the data were obtained that was considered credible. Activities in data analysis to be used in this research were data reduction, display data, and conclusion drawing/verification.

3 RESULTS AND DISCUSSION

3.1 *Strategy to internalize the value of nationalism in tunagrahita students*

Habituation of the value of nationalism can be passed through programs. The prioritization of nationalism education through the process of internalization (dreaming) with habituation approaches needs to be done early on each individual (Anwar, 2014). Programs internalizing the value of nationalism in visually impaired students can be categorized over weeklies, monthly, and yearly programs. Weekly programs include flag ceremonies and scouting. Monthly programs include *kebaya* dress and dance art. Then there is the annual national day commemoration program.

First, the flag ceremony became one of the programs to instill the value of nationalism in visually impaired students. The presence of flag ceremonies can make students with special needs, especially people with disabilities, to be able to participate in the efforts to defend the state. In addition, through the flag ceremony, students honor the services of the heroes who have died in the fight for the independence of the Unitary State of the Republic of Indonesia. Second, scouting is one of the extracurricular activities. All students are required to participate in scout extracurricular activities because this activity aims to shape the spirit of nationalism, self-worth, and also form a good personality in the students. This is in accordance with the form of internalization explained by Wantah (2007) that internalization or planting of value can be done through habituation by doing something repeatedly.

Third, *kebaya* is one of the official clothes of Indonesian women. Female students are required to dress in *kebaya*. It is intended to introduce students to one of Indonesia's female cultures, who often wear *kebaya* to attend a certain event. This is in accordance with the nationalism value indicators described by Aman (2011) quoted by Anggraini (2015) which is proud of its diverse culture. Fourth, dance is one of the arts in Indonesia. Each dance must have different philosophical values and symbolic meanings. Dance art taught to students with special needs becomes a means to develop a positive personality in each student. Through the dance arts program is expected to develop and hone the talents and skills possessed by visually impaired students in particular. In addition, students can also get to know their own culture and not foreign cultures.

Fifth, national day commemoration aims so that students do not forget the history of the Indonesian nation. The history of when heroes died fighting to seize and defend Indonesia's independence. Through this program, students have indirectly remembered the services of heroes who selflessly strived for the welfare of the Indonesian nation. It is in accordance with the purpose of internalization explained by Wantah (2007) that good character is the purpose of the internalization process because in this process one is expected to apply the character that has been included in the internalization process.

Resources are a potential owned by a particular institution, in this case, an educational institution. The potential for an educational institution will be very helpful to achieve an educational goal. Instilling the value of love of the country to the visually impaired can be done by the construction of infrastructure (Huda et al. 2017). The infrastructure is one example of resources. Each institution has different potentials for one another. Some of the resources in question are in the form of technology, financial resources, and human resources.

The regulation, commonly known as school discipline, are one of the means used to support the instilling of nationalism values in visually impaired students. Internalization can be done through continuous strengthening that is expected to motivate students to keep repeating their actions (Saputro et al. 2020). The strengthening is manifested in the form of regulations. The rules for the school of disability students are made slightly different from the rules in schools in general. School of disability students creates disciplines using simple language and is supported by the use of attractive designs so that students are interested in reading them and can understand them easily.

The strategy of internalizing the value of nationalism in visually impaired students has been carried out through optimal teaching and learning activities in the classroom. It is based on the fact that internalization is essentially an effort to share knowledge, which is interpreted as methods, procedures, and techniques in the knowledge management cycle used by educators to share their knowledge with students, in this case between teachers and students (Mufti, 2018). Optimal learning can be achieved through the use of various methods and learning media that are interesting and in accordance with the character of the students.

3.2 *Obstacles faced in internalization of nationalism value in tunagrahita students*

In its implementation, strategies to internalize the value of nationalism to developmentally delayed students faced obstacles. These constraints certainly need to be analyzed further, in order to be resolved immediately so that the internalization process runs smoothly. The obstacles that have been faced in internalizing the value of nationalism in visually impaired students, including the lack of support from parents, differences in character in students, and the difficulty of finding teachers who are in accordance with school standards as an intermediary to instill nationalism values in visually impaired students. This is in accordance with the internalization process described by Wantah (2007) influenced by various factors, namely family, school, and student ability factors.

The lack of support from parents is one of the obstacles that hinder the process of internalizing the value of nationalism in visually impaired students. Success cannot be separated from accuracy during the mentoring process (Muhajir, 2018). For children with special needs, parents become the determinant as well as role models because everyday students spend a lot of time at home. The school is just the second environment to support the success of the internalization of nationalism values in students. If in school students have been taught several theories or practices related to the implementation of the value of nationalism then at home with parental guidance, students must relearn all the learning that has been taught in the school so that students can understand the maximum. If this is not done, then it can be said that learning in school will be useless.

The difference in character that each student has, becomes an obstacle experienced by teachers in teaching and learning activities in the process of planting the value of nationalism. The class system, according to Wihyanti et al. (2018), has been formed as a form of socialization efforts that inclusion and internalize togetherness as unity between friends in the diversity of backgrounds. Students who are in the same class, do not necessarily have the same character. Teaching students with special needs is different from students in schools in general that can be likened to students from each other. Learning done on students with special needs must adjust to the abilities of the students, even if the students are in the same class. Teachers must use a variety of techniques and strategies in learning so that the expected goals can be conveyed to students.

Teachers who teach students with special needs must have more skills than teachers in general. The school finds it difficult to find teachers who meet the standards applied by the school, as not everyone is willing to teach students with special needs that are different from the students in general. Teaching students with special needs is not as easy as doing teaching and learning activities with normal students. The internalization process can be experienced directly by the student when a teacher informs the good and bad of things (Riska, 2020). The teacher of a student with special needs must have more patience, in the face of all the characters that the student has. This makes one have to think twice about teaching students with special needs. That is one of the reasons that makes it difficult for the school to find teachers who meet the standards that have been determined. These constraints must be addressed immediately so that the internalization of the value of nationalism in visually impaired students can run smoothly and produce maximum results.

3.3 *Efforts to overcome constraints in the internalization of the value of nationalism in tunagrahita students*

Efforts to overcome constraints on the internalization of the value of nationalism in visually impaired students have been planned and implemented taking into account all aspects that will be affected. These efforts include providing insights to parents, memorizing the character that each student has, and increasing the creativity of teachers. This is in accordance with the internalization process, namely the identification described by Rais (2012) as quoted by Wardani (2019) that internalization is faster realized through the involvement of role-models.

Giving advice becomes an effort that is realized in the form of insights to parents. The provision of insights to parents is carried out as one of the efforts that will be made by Special School to overcome one of the constraints on internalizing the value of nationalism in visually impaired students related to the lack of support from parents. This is one of the strategies to embed the value of nationalism in learning, one of which is done by giving advice to students (Supriadi et al. 2014). The insights given to parents in the form of advice and direction for parents to support the school in instilling the value of nationalism in students. In addition, parents are also given an idea to keep an eye on their child's future of their shortcomings. Students with special needs also have potential and if guided continuously this will be beneficial to their lives in the future. Therefore, parents of students with special needs should not despair because they know the condition of their children who are not like students in general. Behind the shortcomings of course there are still advantages that need to be developed.

Every student has visually impaired students who have different characters of each other. This requires a teacher to use different methods or media on each student in the process of internalizing the value of nationalism. A child tends to always be active, dynamic, and have a high curiosity. Success in the world of children's education can be achieved through the use of various methods that allow children to follow the learning well (Aisyah et al. 2018). Efforts have been made by the teacher to overcome these obstacles by memorizing the character of each student in order to choose the right method and media.

The school has its own efforts to overcome the obstacles faced, namely by increasing the creativity of existing teachers. Increasing teacher creativity is a temporary effort that can overcome the difficulties faced by schools to find teachers who meet the standards. Indonesia has a pluralistic society, which is often hit by social conflicts, it takes citizens with strong personal characteristics and can live functionally in times of highly competitive globalization (Maftuh, 2008). Efforts to increase the creativity of teachers are carried out by inviting all teachers to participate in seminar activities, either online or offline. The seminars that were attended were certainly related to the ways that can be used by outstanding school teachers in order to be able to carry out optimal learning. The more frequently attending seminars, the teachers will gain a variety of new knowledge that can be applied in daily learning even though the teachers basically have different areas of expertise.

4 CONCLUSIONS

The strategy of internalizing the value of nationalism in visually impaired students was carried out using four kinds of strategies, namely through programs, resource utilization, policies and regulations, and optimal learning. In its implementation, strategies to internalize the value of nationalism to mentally retarded students faced obstacles. They were a lack of support from parents, differences in character in students, and the difficulty of finding teachers who are in accordance with school standards. Efforts to overcome obstacles were providing insights to parents, memorizing the character that each student has, and increasing teacher creativity.

REFERENCES

Aisyah, E.N. et al. 2018. Internalisasi nilai karakter nasionalisme melalui dongeng dan tari (DORI) bagi anak usia dini. *Golden Age: Jurnal Pendidikan Anak Usia Dini* 2(2): 26–34.

Anggraini, S.D.A. 2015. *Peran museum brawijaya dalam menanamkan nilai-nilai kepahlawanan dan nasionalisme terhadap masyarakat.* (Unpublished Thesis). Malang: FIS UM.

Anwar, C. 2014. Internalisasi semangat nasionalisme melalui pendekatan habituasi: Perspektif filsafat pendidikan. *Analisis: Jurnal Studi Keislaman* 14(1): 159–172.

Ardianto, A. 2013. Praktik sosial anak berkebutuhan khusus yang mengikuti *behaviour therapy*: Studi kasus pada anak penyandang autisme di Surabaya. *Paradigma* 1(1): 5–6.

Huda, F. et al. 2017. Penanaman nilai cinta tanah air terhadap penyandang tunagrahita (toleransi terhadap warga kampung idiot Desa Karangpatihan dalam menjaga kebersihan diri dan lingkungan. *Ngabari: Jurnal Studi Islam dan Sosial* 10(2): 50–66.

Maftuh, B. 2008. Internalisasi nilai-nilai Pancasila dan nasionalisme melalui Pendidikan Kewarganegaraan. *Educationist* 2(2). 134–144.

Moleong, L.J. 2016. *Metodologi penelitian kualitatif.* Bandung: Remaja Rosdakarya.

Mufti, D.F. 2018. Internalisasi nilai nasionalisme mahasiswa melalui organisasi ekstra kampus: Deskriptif pada pergerakan mahasiswa Islam Indonesia cabang Tulungagung. *Jurnal Rontal Keilmuan PKn* 4(2): 57–64.

Muhajir, M.M.K. 2018. Penanaman nilai nasionalisme melalui budaya sekolah bagi peserta didik berkebutuhan khusus di SMP Negeri 3 Sidayu Gresik. *Kajian Moral dan Kewarganegaraan* 6(2): 807–821.

Rahayu, F.D.N. 2019. *Pembinaan karakter nasionalis pada siswa tunagrahita di SLB Pelita Qolbu Donomulyo Kabupaten Malang.* (Unpublished Thesis). Malang: FIS UM.

Riska, D.F. 2020. Internalisasi nilai-nilai nasionalisme dalam pembelajaran PPKn di Madrasah Ibtidaiyah Maarif Condro Jember. *Educare: Journal of Primary Education* 1(2) 207–220.

Saputro, R.D. et al. 2020. Internalisasi nilai-nilai nasionalisme dalam pembelajaran Pendidikan Pancasila dan Kewarganegaraan: Studi pada siswa kelas X SMA Negeri Gondangrejo. *Jurnal PPKn* 8(1): 81–90.

Sugiyono. 2015. *Metode penelitian pendidikan (Pendekatan kuantitatif, kualitatif dan R&D).* Bandung: Alfabeta.

Supriadi, A. et al. 2014. Internalisasi nilai nasionalisme dalam pembelajaran PKn pada siswa MAN 2 Model Banjarmasin. *Jurnal Pendidikan Kewarganegaraan* 4(8): 649–655.

Wantah, M.J. 2007. *Pengembangan kemandirian anak tunagrahita Mampu Latih.* Bandung: Ministry of National Education.

Wardani. 2019. Internalisasi nilai dan konsep sosialisasi budaya dalam menjunjung sikap persatuan masyarakat Desa Pancasila. *Nusantara: Jurnal Ilmu Pengetahuan Sosial* 6(2): 164–174.

Wena, M. 2012. *Strategi pembelajaran inovatif kontemporer: Suatu tinjauan konseptual operasional*. Jakarta: Bumi Aksara.

Wihyanti, R. et al. 2018. Internalisasi karakter nasionalisme dalam kediversitasan etnis di Sekolah Dasar Islam. *Edukasia: Jurnal Penelitian Pendidikan Islam* 13(1):79–104.

Empowering Civil Society in the Industrial Revolution 4.0 – Sukowiyono et al (Eds)

Moral coaching of street children in the used prostitution area

M.G. Sadjali*, S. Awaliyah & A.R.A. Atok
Universitas Negeri Malang, Malang, Indonesia

I.W.P. Utami
Nanyang Technological University, Singapore

ABSTRACT: This research is based on the importance of moral coaching for street children in the used area localization. The tendency of negative behavior is carried out so that the negative stigma is attached to the identity of the street children of Balong Cangkring. This research uses a qualitative approach to take descriptive research type. Sources of data in this study are primary data and secondary data. Collecting data using participatory observation techniques, in-depth interviews, and literature study. The analysis of this research uses data reduction, data presentation, and withdrawal. The results of this study focus on moral coaching programs which include, morals instilled in street children, barriers to moral coaching and efforts to overcome obstacles to moral coaching.

Keywords: moral coaching, street children, area localization

1 INTRODUCTION

The number of street children in the used area localization reached 95 (2020) assisted children, there are 45 active children and 50 passive assisted children. The difference between active and passive street children is seen in the level of participation in participating moral coaching programs. Participation in following coaching is more actively carried out by active assisted street children because self-motivation is starting to form while passive assisted street children still often spend their time on the streets and only take part in coaching activities on certain events.

The behavior of actively assisted children has led to positive behavior even though there are still negative behaviors such as smoking, taking to the streets, and skipping school. Personality politeness and language ethics have shown politeness ethics. This is because the self-concept of actively assisted street children has led to a positive direction. The higher the self-concept, the lower the delinquency because they can see, understand, and understand themselves both advantages and disadvantages (Riyadi, 2016). When people are able to see themselves positively, they will try to improve their nature and trust so that they can place themselves and be able to adjust to the norms in society.

Passive fostered children are classified as naughty, unethical children, and often make noise. Negative behavior is often carried out such as: saying dirty words, dishonesty, smoking, hanging out with gambling players in the Balong Cangkring environment, and experiencing school dropouts. The relatively low intensity of coaching activity can even be said to have never participated in coaching activities which will have an impact on the habituating of negative behaviors in passive assisted children who are still attached to them.

*Corresponding Author

DOI 10.1201/9781003180128-13

The young Majapahit movement community is a youth association formed on the basis of common goals and organized territorial relations within the scope of regional unity in Mojokerto City. The presence of the young Majapahit movement community provides moral guidance and skills training to street children of Balong Cangkring. Members of the young Majapahit movement community consist of various alliances such as community volunteers, university students and high school/vocational high school students.

2 METHODS

This study used a qualitative approach, the researcher collected detailed information about the moral development of street children in the young Majapahit movement community. This type of research is descriptive research. The qualitative approach is an approach based on the acquisition of qualitative data then interpreted the findings in the field (Sugiyono, 2014). Primary data sources are obtained from: the community leader, Ainur Rochmatin Fitriani, the core committee, and 8 active assisted children. The secondary data source is in the form of relevant literature in the form of books, scientific journals, these and other secondary data in the form of photos of coaching activities. Data collection using participatory observation techniques, in-depth interviews, and document study. After the data is collected, the data is analyzed using data analysis techniques from Miles and Huberman, whose rules include data reduction, data presentation, levers, and drawing conclusions (Sugiyono, 2014). Data validity was performance during the data collection process, through triangulation of sources, and observer persistence.

3 RESULTS AND DISCUSSION

The young Majapahit movement community was formed in 2016, in 2013 it was known as the SSC (Save Street and Children). The visions this community to helps equalize an independent life to prosper and ease problems social activities for street children. The aim to improving the qualities of life for street children for the better and encouraging them to continue doing teaching and learning activities at school; practicing self-discipline.

3.1 *Moral coaching program in the Young Majapahit movement community*

The moral coaching program for street children in Balong Cangkring includes a social coaching program, social welfare program, and academic education program. Social coaching programs are focused on cultivating a sense of empathy. This sense of empathy is manifest by cultivating social care to help one another. The coaches in the young Majapahit movement community apply it in fundraising activities. Balong Cangkring street children have contributed several times to victims of natural disasters, one of which is the earthquake victims in Donggala Palu.

Even though the condition of the assisted children still needs help, the Majapahit youth movement community tries to make them have a spirit of social care so they don't expect gifts from others, but the assisted children can also give to others according to their wishes. Fundraising activities as an effort to form sympathy for the conditions felt by others. That sympathy is instilled in the street children of Balong Cangkring by itself when they see people who need help, their conscience will be moved to provide assistance.

Social welfare programs are also provided for street children with the aim of equalizing to get non-cash social assistance. The social welfare program is a form of equalization of life for street children as written in the mission of the young Majapahit movement community to provide welfare for the life needs of the assisted children. Compensation activities are carried out to improve the standard of living, fulfill school fees, provide food assistance, and school needs by providing cash and non-cash assistance, usually compensation provided by donors and the community to participate in the assisted community (Suryadi & Zanah, 2016). Social welfare programs are realized through the following programs: blessing spending program and mass circumcision program.

The blessing spending program is an annual program that is carried out once a year in the young Majapahit movement community, the program is a form of community service to improve equality and welfare for assisted children. The activity of the blessing shopping program is to buy basic necessities such as groceries based on the obtaining of vouchers that have been given by the core committee in the young Majapahit movement community. The implementation of the gift shopping program is carried out by collecting as many points as possible when coaching children carry out coaching activities in the young Majapahit movement community every Saturday and Sunday. As many points as possible are collected each week to get shopping vouchers with a maximum total of IDR, 250,000 shopping vouchers. Approval points for assisted children to follow coaching are active, tertiary, not noisy and disciplined as a form of appreciation for good behavior habits carried out by the assisted children, total points will be accumulated at the end of each month of Ramadan.

The mass circumcision program is an annual program implemented in the young Majapahit movement community, for its own implementation in collaboration with several related parties such as foundations, Al-Quran Learning Centers (TPQ), and Islamic boarding schools. Collaboration is most often done with foundations and donors who have the desire to circumcise the fostered children of the young Majapahit movement community for free. Mass circumcision is included as an annual program aimed at circumcising younger brothers in the community of the young Majapahit movement who are old enough to be circumcised and their goal is to perfect their faith.

The academic competence of the fostered children is also one of the teachers' priorities to foster the assisted children in the community to have excellence in the academic field. The assisted children often receive a warning from the class teacher because they are considered less enthusiastic about participating in learning at school. The class teacher informs the community leader so that the fostered children are motivated and given strengthened knowledge in the academic field while participating in moral coaching to be more active in studying at school.

Academic education programs are developed through mathematics classes, language classes, science practicum, studying astronomy, and sex education. Tutoring activities, carried out by providing formal lesson assistance to improve the academics of the fostered children, besides that the purpose of tutoring activities is to motivate the assisted children to continue their education at the next level of education (Suryadi & Zanah, 2016). Tutoring can help the cognitive, affective, and psychomotor coaching of street children in Balong Cangkring to improve their competence for attainment in formal education.

The sex education program has become a superior fostered program implemented as an effort to prevent sexual crimes and provide prevention children from being affected by acts of social deviation considering that the environment where the assisted children in the young Majapahit movement community live is in a former localization area. The possibility to be affected is very large, because the environment as a place to interact always brings good and bad impacts if the individual does not have an understanding that can be used as a guide for assessing his own personal behavior and can fall into negative actions. According to Hayati (2019), the environment with bad habits provides a stimulus to the child's mind to imitate behavior carried out in the environment, the imitation process becomes the beginning of social behavior deviation so it takes time to change these bad habits through a gradual coaching process.

The handicraft program that was developed was hand puppet craft, with the aim of improving skills and changing mindsets from an early age to develop valuable handicrafts to cut activities on the streets. Craft training can improve skills to develop creativity. Creativity can develop through training to make handicraft products, with a craft program, assisted children can gain learning experiences so that they can continue to develop to make handicraft products that have selling value (Abdul, 2017).

Motivational guidance for street children of Balong Cangkring is very much needed as an effort of social assistance and control. Immature moral development must be monitored through guidance efforts, the transformative learning model is selected as a method of motivational guidance. The transformative learning model can give positive changes for parents and children, parents can give attention and affection to the child so that the child's mental condition is getting better, such as self-confidence begins, and spiritual quality is getting better. The transformative learning model of moral coaching is carried out by: (a) changing the mindset of parents through awareness; (b) change the mindset of parents; (c) providing motivation; (d) spiritual support; and (e) protection (Nusantara & Moedzakir, 2015).

3.2 *The morals to plant in Balong Cangkring street children*

The behavior of street children in Balong Cangkring is still based on the fear of punishment. Kohlberg's theory of moral development it is explained that pre-conventional moral reasoning is based on objects outside the individual as a measure of right and wrong, while obedient orientation is based on fear of punishment, a behavior is considered right if not punished and wrong when it needs to be punished (Adisusilo, 2016). If the theory of moral development is correlated with the moral development of street children in Balong Cangkring, it can be said that the morality of street children in Balong Cangkring has not been properly formed so that moral cultivation is needed.

The immoral behavior which is still used by street children in Balong Cangkring is the personal cause of the assisted children who cannot be said to be moral. There are several street children in Balong Cangkring at the age that are still considered their children have problems because of cases of sexual harassment. This results in them getting a stigma that is not good in the community and friendship. Immoral behavior that is still used to it causes the negative stigma of street children to stay. So that a coaching process is still needed, here are 4 moral focuses instilled in the street children of Balong Cangkring: moral relation to god, to human, to environment, and to self.

Moral cultivation in relation to God is a form of teaching that comes from religious norms. Religious norms are sourced from the teachings of God almighty, which has religious principles that can be used as a guide for life, the teachings of goodness in religion to carry out the good commandments in relation to God, fellow humans, and the environment become guidelines for goodness to be done by every human being (Wardhani, 2015). The devotion of every human being to religion can be seen from their faith in carrying out orders and obligations according to their religion and beliefs. Religious understanding is important because in religion there are rules regarding how a person should behave so that religion becomes the guide and guide for every human being, any religion that is believed to be true by the adherents of these religions must teach goodness to strengthen the personality of the fostered children in the young Majapahit community religious guidance as a provision for street children in Balong Cangkring to live in society.

Moral coaching is related to God for the fostered children in the young Majapahit movement community as a means of forming attitudes, mental spirituality, and understanding religious life in the implementation of daily life. Coaching activities are carried out by providing concrete examples to assist children to facilitate the coaching process. Direct practice regarding religious moral fostering activities can be demonstrated through good habits by providing direct examples of performing prayers and reading prayers correctly (Agustin & Widodo, 2018). The process of giving concrete examples to assisted children by providing positive stimuli so that the assisted children can receive examples of the habit of carrying out prayers properly. Spiritual support is carried out through the cultivation of religious values as a foundation to increase piety to God Almighty and deepening religious teachings according to beliefs (Nusantara & Moedzakir, 2015).

Cultivating moral relations with fellow human beings (social) is a manifestation of the norms of decency (custom). The norm of decency regulates behavior based on customs, culture and tradition regulations, the source of the norms of decency, namely customs, traditions or culture, so that good habits in the area are not necessarily considered moral in other areas because each region has a guideline for norms of decency which cannot be considered the same or universal (Wardhani, 2015). Moral cultivation in relation to fellow human beings (social) is instilled through courtesy. The guidance given is in the form of mentoring and examples such as language ethics, honesty, politeness and discipline. The coaching model chosen is the home visit. Home visit guidance is carried out through an approach activity in the form of learning assistance by forming small groups teaching politeness ethics related to character building (Amin, et al. 2014).

Moral cultivation in relation to the environment emphasizes gratitude and love for the environment. Environmental awareness is instilled through community service activities such as sweeping, dumping garbage in its place, and watering plants. Community service is part of the cultivation values related to respect for the natural environment. According to Itariyani (2013) community service activities are not only carried out through sweeping and cleaning the yard but also activities to protect plants in the surrounding environment so that the surrounding environment remains green and fresh so that it can help environmental health and have an impact on environmental health for human life.

3.3 *Barriers to moral coaching in the Majapahit Youth movement community*

There are inhibiting factors for the implementation of moral coaching in the young Majapahit movement community, include: infrastructure provision is the main problem due to the lack of budget funds. According to Hayati (2019), the last budget for funds can affect the ability to give facilities, both provisions of facilities and infrastructure, such as: difficulty in bringing in coaches from outside who are competent in the field of moral coaching, and cannot give a place for guidance. Limited infrastructure has an impact on the relatively short time of guidance given, due to agreement with Elementary School (Mentikan 6).

Participation in volunteer activity is also an obstacle to coaching, so that the collaboration between the two is not harmonious. This will have an impact on the ongoing development. Volunteer harmony affects motivation and interest, if volunteer treatment of street children is warmer, such as their own family, it can make the second relationship closer (Ulfah, 2019).

Acceptance parental support results in the behavior of street children in Balong Cangkring who tend to prefer to do work on the streets than to participate in coaching activities because of the motivation of parents' orientation towards children more directing to work on the streets. This habit will make children have low motivation and be orient to continue doing work on the streets in the end have an impact on the child's personality who is carried away with the environmental conditions of street children. A person's personality on an object that is built by the values of his life background, the existence of a past linkage will form these values in his personality or mentality so that he cares about values into his belief (Muhtadi, 2011).

4 CONCLUSIONS

Based on the results of research and discussion, it can be concluded that the moral coaching of street children in Balong Cangkring is still not good. This is evidenced by the immoral behavior carried out by street children of Balong Cangkring. The habit of behaving in accordance with the norms in society really needs to be re-instilled through the process of cultivating moral relations with God, on fellow humans (socially), on oneself, and relating to the environment. To get maximum coaching results, of course, it must be developed through educational coaching programs such as moral development programs in the young Majapahit movement community which are not only focused on social development programs and educational coaching, there are several supporting moral development programs such as: social welfare programs, craft programs and training, as well as motivational guidance programs.

REFERENCES

Abdul, P.M. 2017. Peran pekerja sosial dalam meningkatkan kemandirian anak jalanan melalui pelatihan melukis di UPTD kampung anak negeri Wonorejo Surabaya. *Jurnal Pendidikan Luar Sekolah* 6(2): 1–10.

Agustin, A.D. & Widodo, R. 2018. Model pembinaan anak jalanan di Pondok Pesantren Salafiyah Sabilul Hikmah Malang. *Jurnal Civic Hukum* 3(1): 1–8.

Adisusilo, S. 2016. *Pembelajaran nilai karakter konstruktivisme dan VCT sebagai inovasi pendekatan pembelajaran aktif.* Jakarta: PT Rajagrafindo Persada.

Amin, M. A. et al. 2014. Pelayanan sosial bagi anak jalanan ditinjau dari perspektif pekerjaan sosial. *Prosiding Penelitian Dan Pengabdian Kepada Masyarakat* 4(2): 1–9.
Hayati, M. 2019. Pembinaan moral keagamaan dan implikasinya terhadap perubahan perilaku narapidana di blok melati LP Kelas II A kota Mataram. *Jurnal Pendidikan Guru Madrasah Ibtidaiyah* 44(1): 1–13.
Itariyani, N. 2013. *Pembinaan moral pada remaja putus sekolah di balai rehabilitasi sosial "Wira Adhi Karya" Ungaran.* Universitas Negeri Semarang.
Muhtadi, A. 2011. pengembangan sikap dan perilaku siswa yang bermoral dalam kegiatan pembelajaran di sekolah. *Majalah Ilmiah Pembelajaran* 7(1): 1–12.
Nusantara, W. & Moedzakir, M. 2015. Pembelajaran transformatif pada kegiatan pendampingan anak jalanan di kota Malang.*Jurnal Pendidikan dan Pembelajaran* 22(1): 1–13.
Riyadi, A. 2016. Hubungan konsep diri dengan kenakalan anak jalanan pada rumah singgah putra mandiri Semarang. Psympathic: *Jurnal Ilmiah Psikologi* 3(1): 1–12.
Suryadi & Zanah, Y.M. 2016. Pemberdayaan anak jalanan berbasis komunitas. *Jurnal Dakwah Dan Komunikasi* 4(1): 1–15.
Sugiyono. 2014. *Metode penelitian kuantitatif, kualitatif, dan R&D.* Bandung: Alfabeta.
Ulfah, F. 2019. *Strategi rumah pintar bangjo dalam pengentasan anak jalanan di daerah pasar johar kauman semarang.* Universitas Negeri Semarang.
Wardhani, P.S.N. 2015. *Pembinaan moral remaja di Balai Rehabilitasi Sosial Mandiri Semarang II Kota Semarang.* Universitas Negeri Semarang.

Empowering Civil Society in the Industrial Revolution 4.0 – Sukowiyono et al (Eds)

Full day school program in forming environmental care character of junior high school students

T. Uda*, E. Karliani, Sakman & Dotrimensi
Palangka Raya University, Palangka Raya, Indonesia

ABSTRACT: This study aimed to analyze and evaluate the implementation of full-day schools with the character of environment care. This study was a qualitative descriptive, that was analyzed through descriptive techniques, specifically reduction, display, presentation, and conclusion. The results are: (1) readiness of facilities and infrastructure is classrooms, parking, sports facilities, worship facilities, student playground, gardens or parks, canteens; (2) programs run at school environment care characters are class pickets, *musa* (picking the trash), *jumsih* (Friday cleaning), hydroponic parks, outdoor activities, reducing the use of plastic waste, and class cleanliness competitions; (3) the implementation of the full-day school has allowed students to feel happy because they can interact longer with their friends and can do many positive activities such as being involved in watering plants, caring for their hygiene. Parents agree because it can increase the formation of character, especially religious and social character.

Keywords: full day school, environment care, character

1 INTRODUCTION

The environment care character is one of the 18 characters developed by the Ministry of Education and Culture. It is related to how students have an understanding, attitude, and behavior that loves to take part in protecting the environment around their homes for their survival and future generations. Every human life depends on the environment. A clean, healthy environment is the right of every. There are several examples of students' behaviors that do not reflect their concern for the environment, one example is the behavior of littering, damaging the green plants of the school, polluting the walls of the school, and lack of caring for classrooms and school's cleanliness in general. Palangka Raya City is one of the cities which has local characteristics which are located in river basins and peat swamp areas. Environmental care is needed to prevent disasters caused by human activities on the environment, one of which is forest fires and floods (Suparwadi, 2013; Andikurrahman, 2012).

Several characters have been studied in previous studies as the result of full-day school activities. They have produced findings that the implementation of full-day school activities will improve students' religious character, increase learning motivation, and improve students' critical thinking skills. Meanwhile, no research examines the implementation of full-day school about the environmental care character for State Junior High School/MTs students. Considering that one of the problems in Palangka Raya City is related to the environment, the results of research that become the community's concern for the environment in Palangka Raya City are about waste management.

The behavior of the people of Palangka Raya City in disposing of garbage is not good and orderly. For example, the garbage disposal schedule that has been set by the City Government through the cleanliness and city planning service has not been implemented properly, there is still

*Corresponding Author

DOI 10.1201/9781003180128-14

the behavior of the people who throw garbage not at the garbage bin but into rivers and forests that are around the residents' settlements, there are still many forests which is near the highway which is used as a landfill so that garbage is scattered into the streets (Suparwadi, 2013; Sutarmin, 2014; Alhamuddin & Buchori, 2016).

This study needs to be done to analyze the implementation of a full-day school program, and provide alternative solutions to problem-solving by developing a model of full-day school activity programs to form the character of environmental care. The question in this study how the implementation of full-day school can form the character of caring for the environment if the learning activities at school provide understanding, awareness, and habitual behavior to students to love the environment and have care (respect) for the environment?

2 METHODS

This study is a qualitative descriptive study to analyze and evaluate full-day school programs informing the environmental care character of junior high school students in Palangka Raya City. The stages of the research will be carried out and elaborated as follows: (1) the research team developed a research instrument based on the indicators of the full-day school program informing the environmental care character of junior high school students in Palangka Raya City. The indicators of implementing a full-day school program include readiness of school infrastructure facilities in the implementation of full-day school programs to form the character of environment care (checklist list); activities program in implementing full-day school to form the character of environment care; human resources readiness (principals, teachers, non-academic staffs) in implementing full-day school programs to form the character of environment care; readiness of the students in implementing full-day school programs to form the character of environment care; parental support in implementing full-day school programs to form the character of environment care. (2) Gathering the data in the schools that have implemented a full-day school program with techniques: surveys, interviews, observations, and documentation. (3) The team conducted data analysis to conclude the full-day school program to form the environmental care character of State Junior High School/MTs students in Palangka Raya city following the data collection techniques. The results of the study were analyzed using qualitative descriptive data analysis. (4) Based on the results of the field data, the team compiled a report related to research and formulated several full-day school program activities to be implemented by the school.

To find out the successful implementation of the full-day school program, it was carried out in the form of an interview guide in the form of an open question list for the school principal, teachers, students' parents, and the students themselves related to environmental care character indicators, specifically: (1) always maintain environmental preservation, (2) do not take cut down or uproot vegetation along the way, (3) do not scribble, write inscriptions on trees, stones, roads or walls, (4) always throw trash in its place, (5) do not burn rubbish around housing, (6) carry out environmental cleaning activities, (7) collect used goods, and (8) clean up trash that clogs waterways.

3 RESULTS AND DISCUSSION

The research team determined the research locations were in SMPN 1 Palangka Raya, SMPN 2 Palangka Raya, SMP Christian Golden School, SMP IT Hasanka. The selection of research locations was based on the representation of the public junior high school based on the observations of the researchers having readiness in conducting a full day school research is as follows:

3.1 *Readiness of the school infrastructure facilities in implementing full-day school programs to form the character of environment care*

SMPN 1 Palangka Raya has a school corner for subjects, especially Pancasila and Civic Education (PPKn) and Bahasa Indonesia, for the future the school will develop a school corner for other subjects. The school corner is designed to strengthen the achievement of subject competencies in addition to the activities carried out in the classroom. In the corner of the school students can learn while breathing fresh air, outside the room by learning everything around them. In addition, SMPN 1 Palangka Raya also has a clean canteen that provides food for students, has a school garden, hasKa trash can, hygiene advice in the classroom and outside the classroom, and has live plants in the classroom.

SMP IT Hasanka has a distinguished student rest area for both boys and girls which has a clean canteen that provides food for the students, has a school park, has trash cans, sanitary facilities in the classroom and outside the classroom, and have live plants in the classroom. At SMP Golden Christian School, the school provides a school canteen, which provides healthy food catering for students. They can choose the food they want according to the menus provided. During breaks, students are required to eat in the dining room, not in the classroom, and they can play in the park and school environment. They must not be discharged from school when it is still school hours. SMPN 2 Palangka Raya already has a green park and a school garden, a sports field, has a clean canteen that provides food

The readiness of school facilities and infrastructure to support the implementation of full school can be run well. The availability of school facilities and infrastructures such as classrooms, parking lots, sports facilities, worship facilities, student playgrounds, gardens or school parks, a canteen that adequately provides the food needs of students, will provide space and a comfortable feeling while students carry out full-day school activities in school because they will be in school for about 8 hours or more. In implementing full-day school, it should be done with various activities that take place actively, creatively and fun. The full-day school system is more possible for the realization of the effectiveness of the learning process and improving the quality of learning (Hasan, 2006; Astuti, 2013; Sutarmin, 2014).

3.2 *The program of implementing the full day school to form the character of environment care*

The implementation of full-day schools in several schools is carried out with additional activities that are deliberately designed to fill school hours after teaching and learning activities are carried out. In SMPN 1 Pangka Raya, the program was carried out with several activities, namely co-curricular activities: each class has work (making light trees from the waste, light trees from corn husks). These activities are included in the subjects of crafts, flag ceremonies, literary movements: there are students' literacy data, in the literacy movement the principle directs the activities carried out in the literacy activities. The literacy movement can be done by making posters, sticking formulas in classrooms, installing the waste that has become handicrafts in classrooms. There is a green school program (green constitution), DLH Excellent achievement, leaving students on time implicitly in the classroom and outside the classroom, the teacher supervises on time habits, Friday prayers *(Duha),* worship, teachers also have worshipped every month, namely routine recitation in each class.

The full-day school is applied through extracurricular activities namely in the field of nineteen extracurricular activities such as basketball, volleyball, futsal, silat, karate, table tennis, badminton, flag raisers, teen red cross, scout, choir, drum band, Christianity, Mathematics, Science, English, painting, local dance, Islamic religion *(habsy, tilawah, BTQ)*. In the classroom, there is also a picket schedule that has been agreed by all students to be carried out and if they are violated, penalties are given.

At SMP IT Hasanka, the process of using recycled paper is one of the first steps for the school to reduce environmental pollution because every time we print, we use used paper, and from now on there has been the use of plastic waste in recycling. Every month, we hold an award or reward for the cleanest class as an appreciation and enthusiasm for the class to maintain environmental cleanliness and care, as well as through extracurricular activities that develop students' interests. To face these obstacles, we hold an environmental care activity every Saturday, for the next morning we hold gymnastics clean up then clean the gutters. We watch them clean the trash in the gutter. If there are students who violate the school rules of cleanliness and the others, we solve it through the homeroom teacher first, then the teacher will follow it up with the student and if it is still unable to be solved, then we call the parents.

At SMP Golden Christian School, the school has activities that students must follow. These activities have been discussed in a parenting meeting before starting in the new semester such as assembly, retreat to the beach, natural parks, arboretum, outbound/field trip, Balanga museum, and Taman Wisata. Those activities aim to foster recognition and awareness of the natural environment. The development of school learning processes must be adapted to the material taught in the classroom and related to the methods and facilities and infrastructure provided by the school. While in the school environment, the hydroponic parks are made required by SMP GCS to take care of and care for them. What is planted is a plant for family medicine. The school made pickets to water, cut dead leaves, and clean the environment around the park. In the classroom, there is a picket schedule as well that has been agreed by all students to be carried out and if violated, sanctions are given.

In SMPN 2 Palangka Raya, we conduct the program activities of worship to increase faith and piety; various extracurricular programs, which include more or less extracurricular activities related to sports, spirituality, arts, and other educational fields such as science, English language club besides activities carried out by students, namely the existence of a Friday cleaning program, and cooperation in school hygiene. In the classroom, there is a picket schedule that has been agreed by all students to be carried out and if violated, penalties are given

The full-day school program implemented in all schools which are research sites already have programs in strengthening the character of caring for students' environment, for example in terms of (1) always maintain the flexibility of the surrounding environment, (2) do not take, cut or revoke growth plants found in schools, (3) do not scribble, write inscriptions on trees, stones, school walls, (4) always throw trash in place, (5) do not burn rubbish around the housing, (6) carry out environmental cleaning activities, (7) collect used goods, and (8) clean up the trash that clogs waterways. for example, there are class picket programs, trash movements, *jumsih* (Friday cleaning), hydroponic gardening, outbound to introduce the environment to students, problem-solving rubbish waste for students' creative crafts and creations, reduction in the use of plastic rubbish, and a class cleanliness competition n encourage to have a clean soul and care for the environment of its class. As some of the results of the following interviews explain that there are obstacles in instilling the character of caring for the environment in schools that have implemented full school. Students are given an understanding and accustomed to daily activities to love the environment, for example, there is a school picket program, the movement of watering and caring for plants, making crafts from garbage waste, and the habit of getting used to throwing trash in its place. Usually, on Friday we held gymnastics together, and voluntary work, such as cleaning the TOGA garden (a family medicinal plant), tidying and cleaning the room, tidying bookshelves, etc. (Nenggala, 2007; Taufik, M. 2014; Astuti, 2013; Sutarmin, 2014).

3.3 *Students' readiness in implementing full day school programs to form the character of environment care*

All students agreed with the implementation of FDS in SMP 1 Palangka Raya. They are happy to water every day, there is a duty picket schedule and at the same time the task of watering. In the classroom, some plants are brought by students who must be cared for by each of them. They may not eat in class, and avoid the use of single-use plastics. Students feel happy with full-day school because the teachers also directed and helped flush and clean as well as with some friends who were in picket turns. Every Saturday there is joint community service guided by the clerics. The way they never taught us roughly but more towards inviting gently and never using violence or to hit. They like to water the plants in front of the class. they are taught to help and love the environment as fellow living creatures.

Students feel happy because they have to run a program that has been run by the school can interact longer with their friends at school and can do many positive activities at school. But for school activities that strengthen the character of caring for the environment in general students feel happy if they have to be involved in watering plants, caring for shared hygiene, with examples given by their teachers. Many parents agree with the implementation of full-day school because it can increase the foundation of their character, especially religious characters and social attitudes. The effectiveness of the full-day school in strengthening the character of environmental care is through the habit of clean living in the school environment and at home, as well as an understanding of the habit of littering, taking care, and loving plants for their living environment (Hasan, 2006; Nenggala, 2007; Taufik, M. 2014; Astuti, 2013; Sutarmin, 2014).

4 CONCLUSIONS

Based on research data, several conclusions in this study are: (1) readiness of school facilities and infrastructure implemented by full-day schools is adequate such as classrooms, parking lots, sports facilities, worship facilities, student playgrounds, school gardens or parks, a canteen that is sufficient to provide students with foods, will provide space and support feeling of comfort as long as students carry out full-day school activities in their school because they will be at school for about 8 hours or more; (2) school programs to strengthen students' environmental care character have not involved all elements of students and are simultaneously accustomed to the continual habits of VII to XII graders in high school; (3) the implementation of full-day school has students feel happy because they have to be involved in watering plants, caring for their hygiene, with examples which were given by their teachers. Many parents agree to the implementation of full-day school because it can increase the foundation of the children's character, especially religious characters and social attitudes. The effectiveness of the full-day school in strengthening the character of environmental love is through the habit of living clean in the school environment and at home, as well as an understanding of the habit of littering, taking care, and loving plants for their living environment.

REFERENCES

Alhamuddin. & Buchori. 2016. The effect of multiple intelligence-based instruction on critical thinking of full day islamic elementary school students. *Journal of Islamic Education Ta'dib* 21(1): 31–40.

Andikurrahman. 2012. *Dampak pelaksanaan full day school terhadap perkembangan sosial peserta didik (Studi kasus di SD Plus Nurul Hikmah Pamekasan).* Pamekasan: Department of Tarbiyah, Islamic Religion Study Program STAIN Pamekasan.

Ansari, M.I. 2015. Strategi sistem full day school dalam membentuk empati siswa. *Journal Madrasah Ibtidaiyah Muallimuna* 1(1): 70–78.

Astuti, M. 2013. *Implementasi program full day school sebagai usaha mendorong perkembangan sosial peserta didik TK Unggulan Al-Ya'lu Kota Malang. Jurnal Kebijakan dan Pengembangan Pendidikan* 1(2): 133–140.

Hasan, N. 2006. Fullday school (model alternatif pembelajaran bahasa asing). *Jurnal Pendidikan Islam Tadris* 1(1): 23–31.

Islamika, D. 2011. *Pengaruh full day school terhadap kecerdasan sosial anak kelas IV di SDIT Bina Anak Sholeh Yogyakarta.* (Unpublished Thesis). Yogyakarta: UIN Sunan Kalijaga Yogyakarta.

Ma'ruf, A. 2016. *Implementasi full day school dalam membentuk akhlak siswa di SMP Ta'mirul Islam Surakarta tahun pelajaran 2015/2016.* (Unpublished Thesis). Surakarta: University of Muhammadiyah Surakarta.

Nenggala, A.K. 2007. *Pendidikan jasmani, olahraga, dan kesehatan.* Bandung: Grafindo Media Pratama Publisher.

Nurhayati, I. 2014. Penerapan sistem pembelajaran 'fun & full day school' untuk meningkatkan religiusitas peserta didik di SDIT Al-Islam Kudus. *Jurnal Teknologi Pendidikan dan Pembelajaran, 2*(2): 232–244.

Sutarmin, S. et al. 2014. *Penanaman nilai-nilai dasar humanis religius anak usia dini keluarga perkotaan di TK islam terpadu.* Jurnal Pembangunan Pendidikan Fondasi dan Aplikasi 1(2): 156–166.

Suparwadi, E. et al. 2013. *Analisis dampak peningkatan jumlah penduduk dalam pola perilaku pengelolaan sampah di Kota Palangka Raya.* Joined research with BKKBN Provinsi Kalimantan Tengah.

Taufik, M. et al. 2014. Pengembangan media pembelajaran ipa terpadu berkarakter peduli lingkungan tema 'konservasi' berpendekatan science-edutainment. *Jurnal Pendidikan IPA Indonesia* 3(2): 140–145.

Empowering Civil Society in the Industrial Revolution 4.0 – Sukowiyono et al (Eds)

The implementation of mutual assistance values in a religious day celebration

K. Masita* & Margono
State University of Malang, Malang, Indonesia

A.N. Widiadi
Victoria University of Wellington, New Zealand

ABSTRACT: Mutual assistance had an important role in the plural society life, there was no exception to the people's life in Balun Village, which was known as a village that had religious and cultural diversity. Thus, mutual assistance played a role as social capital in creating tolerance, maintaining harmony, and preventing conflicts between religious communities. This study aimed to describe the implementation of mutual assistance in religious day celebrations in Balun. This study used a qualitative approach, the data was collected by observation, interview, and document study. The results showed that the implementation of mutual assistance in a religious day celebration was done through community service activities, assisting the preparation of celebrations, donating, and cooperating between religious communities in securing during celebrations. The obstacle faced was the community's activities, therefore not all residents could participate in the implementation of mutual assistance. The efforts made were determining the effective time and giving an announcement or socialization related to the activities to be carried out.

Keywords: values, mutual assistance, celebration, religious people

1 INTRODUCTION

Indonesia is an archipelago state that consists of various kinds of ethnic groups, this condition makes Indonesia a pluralistic country. It can be seen as sociocultural and geographic which are so complex, diverse, and extensive with various ethnicities, cultures, customs, languages, religions, and so forth. A plural society according to J. S Furnivall by Rustanto (2016) is a society condition that consists of two or more elements, it could be religion, culture, and ideas that live independently without mixing each other but it is coexistence in a political situation. Talking about pluralistic as the official national motto of Indonesia "Bhinneka Tunggal Ika" in the 1945 Constitution of the Republic of Indonesia Chapter XV Article 36 A that has a meaning "Unity in Diversity," the motto does not only function as a byword but also symbolizes the diversity of Indonesia society, which is united in one unit as the Unitary State of the Republic of Indonesia.

The various conditions of society in a multi-cultural context, on the one side it reflects a large national identity, on the other hand, they also have the potential to cause conflicts that threaten national integration. It can be seen on the characteristics of a plural society according to Pierre L Van De Bergh (in Rustanto, 2016) the division of society into groups that have different sub-culture from one another, has a social structure that is divided into institutions that have non-complementary characteristics, difficult to determine an agreement or consensus among its member towards the

*Corresponding Author

 DOI 10.1201/9781003180128-15

basic values, besides, they also experience conflicts relatively frequent between one group and another, their relative social integration grows on coercion and interdependence in the economy sector, and the existence of political domination by one group over the others.

The plurality of society and the potential of the truth can be seen in the people's life in Balun Village, precisely in Turi District, Lamongan. The Balun society life, which has the requirements for diversity, has made the village known as "Pancasila Village" as the village has religious and cultural diversity. There are three major religions in Balun, i.e., Islam, Christianity, and Hinduism which can coexist peacefully until now. The diversity of religious communities in Balun Village is not the reason for conflicts that generally occur in plural societies. However, it becomes a strength for the Balun Village society in showing the unity and integrity attitude. The harmony exists in the Balun Village society just because of the awareness of mutual respect and mutual assistance between religious communities' life. Mutual assistance is a noble value that is believed that have existed in Indonesian life and it can maintain the integrity and unity of society by helping each other and together in the Indonesian diverse life. As explained by Widiyanti and Sunindhia (2003) mutual assistance is a value system of social life that is from generation to generation in the villages life in Indonesia in the form of concrete social attitude. The advantage of this cooperation is to lighten and strengthen relations between residents and unite the Indonesian people or society.

Mutual assistance value has been a vision of Balun Village society to achieve social welfare, through mutual assistance, the community hopes that the desired goal, namely harmony, and prosperity, can be easily achieved. Thus, mutual assistance in that village is very important and worthy to be kept and preserved. To preserve it, they often did mutual assistance or helping each other that was done during the religious celebration among religious communities. The implementation of each religious celebration cannot be separated from helping and cooperating activities that were carried out by society between the same and different religions who also assisted, helped, and worked together in the process of preparation and execution. Therefore, it is expected that Balun Village could be a pluralistic representation, as an inclusive village that provides a space for the society to implement their belief and culture freely with full tolerance, and could be an example in wider for a pluralistic Indonesian society. Based on the introduction that was explained above, it became a special attraction to have a look deeply at the religious communities' life in the village. Then the research title was formulated as "The Implementation of Mutual Assistance Values in Religious Day Celebration in Balun, Turi, Lamongan".

2 METHODS

The study about the implementation of mutual assistance values in religious day celebration in Balun Village used case study research by describing and explaining the various aspects related to an individual, group or organization comprehensively. This research method aimed to maintain the integrity of the object under study by focusing on a case intensively and detail in general (Ulfatin, 2017). The approach used was qualitative by describing a phenomenon about what research subjects have experienced such as behavior, perception, motivation, and action holistically in the form of words and language at the special natural context and by utilizing various natural methods (Moleong, 2007).

The process of collecting data in this study used observation, interviews, and document study techniques. The data sources that were used consisted of two types, namely primary and secondary data. The primary data in this study consisted of: (1) informants involving Balun Village apparatus, community leaders from each religion in Balun Village and the youths or residents of each religion; (2) events, that was religious holidays celebration; and (3) Documents, those were in the form of village profiles. Whereas the secondary data that was used was from the relevant previous studies or events. In this study, the obtained qualitative data that was conducted from July to September 2020 was analyzed through some steps: (1) processing and preparing data; (2) read the whole data; (3) coding the data; (4) applying a coding process to describe settings, categories and themes; and (5) linking the themes/descriptions (Creswell, 2012).

3 RESULTS AND DISCUSSION

3.1 *General description of Balun Village*

"Balun" is the name of a village which was taken from the name of village elder named Mbah Alun which came from the name Sunan Tawang Alun became Mbah Alun then became Mbalun and finally became Balun. Balun Village is a village located in Turi District, Lamongan Regency, East Java with an area of 621.103 hectares. Balun Village is located in the north of Lamongan Regency, about 5 meters from Lamongan District Government center or city center with a distance of ±15 minutes. Balun Village consists of 2 (two) hamlets, namely Balun and Ngangkrik with a total population of 4.744 people. Most of the areas in Balun Village are rice fields with an area of approximately 530 hectares. It shows that the agricultural sector has an important role in the economic society. So that the majority of people in Balun Village work as farmers. The heterogeneous and pluralistic conditions community make Balun Village as a special village with a harmony that exists among religious communities, namely Islam, Hinduism, and Christianity as well as the culture that develops in the local community life making Balun Village known as "Pancasila Village."

3.2 *The implementation of mutual assistance in religious day celebration in Balun*

The development of various beliefs or religions in Balun Village is not only a set of thoughts that regulate the relationship between humans and God but also the relationship between humans and other humans, and between humans and their natural environment. According to Lubis (2015) the role of religion based on its function consist of two, there are: (1) functions to provide the understanding and explain the viewpoint regarding the world that cannot be reached by humans (beyond), which can lead to significant depression and frustration. Besides, religion also teaches about the awareness of the world view which will come out a work ethic as an ideal reward that a person will receive afterlife (eschatological); (2) religion functions as a ritual that allows human relations with things out of range.

Based on the results of this study, it was known that each religion in Balun Village has different religious activities with various kinds of celebrations. Other religious feast days except during feast celebration tend to be carried out simply and internally, whereas the annual year celebration on a bigger scale is only carried out during the feast. Therefore, in the discussion about the feast day of each religion i.e., Eid al-Fitr as a holiday celebration for Muslims, Nyepi as a holiday celebration for Hindus, and Christmas reception as a holiday celebration for Christians.

Based on each of the implementation of religious holidays, it could be seen that in the celebration of religious holidays in Balun Village, it was found that there were various kinds of implementation of the mutual assistance values. According to the Ministry of Education and Culture (1982) mutual assistance is cooperation in achieving certain goals with the reciprocity principle which creates social activities in society life. Mutual assistance can vary according to the field and social activity. For the implementation, mutual assistance can be spontaneous, it based on a purpose or because they want to fulfill a social obligation for community activities. The various mutual assistance activities in the religious holiday celebration could be grouped by types. According to Bintarto (1980) explained that the forms of mutual assistance are: (1) A mutual assistance in collecting donation of the energy for the common interest; (2) A mutual assistance in collecting donation of thoughts for the common interest; (3) A mutual assistance in collecting donation of funds for the common interest.

The implementation of mutual assistance value in collecting energy in the religious holiday's celebration in Balun Village, it could be seen in the community service activities that carried out during the celebration of Eid al-Fitr, Nyepi, and Christmas. As explained by Nafisah and Sarmini (2020) in their study that mutual assistance in cleaning the village has become a culture for the society which is carried out in the community service activities. Besides commemorating Independence Day, community service activities are also often carried out on big holidays such as before the feast which is carried out by cleaning the tombs. Thus, community service has become a mutual assistance tradition for society.

In addition, the community service in the form of manpower assistance is also carried out by Balun Village society by helping the process of safeguarding and orderliness of the implementation of the feast in each religion. The assistance is not only carried out by the fellow religious members but also with people of different religions assistance and cooperation. It is in line with the study conducted by Rahmayanti (2017) explained that the religious days celebration of all religions; Islam, Hindu, and Cristian take turns in maintaining security and order during religious celebrations. Therefore, between Muslims and Hindus or Cristian and vice versa, they work together, assisted by the authorities to maintain environmental security.

According to Sastropoetro participation was done in reaching the goals by showing the involvement either spontaneously or accompanied by awareness and responsibility for the interest of the group (in Isbandi, 2007). Several forms of community participation are participation in the form of force, participation in the form of funds, participation in the form of material, and participation in the form of information. While the implementation of mutual assistance values in the form of collecting funds voluntarily was seen in the celebration of Eid al-Fitr, it was shown by the participation of the society in Balun Village who made contributions or gave donations voluntary which were used to fulfill all the needs of the celebration. This is in line with the study of Wadu et al. (2019) stated that activities to help each other in community life that aim to advance welfare is also done by giving donations either giving money or things without expecting anything in return or voluntarily. So, it is not only giving force and funds, mutual assistance value in feast celebration can also be implemented in participating in thought which is manifested in the activity of making *ogoh-ogoh* statues in Nyepi celebration in Balun Village, not only Hindus are involved but also other religions people help it.

3.3 *The obstacle faced and the efforts made to overcome in implementing mutual assistance value in the religious day in Balun*

The obstacle faced in implementing mutual assistance value in the religious day in Balun Village was the community's activities, it showed that not all citizens could follow or help the implementation of mutual assistance. The constraint was caused since some people still worked and some of them still conduct their study for the youth who usually helped and participated in implementing mutual assistance for religious holidays celebration. According to Rismayanto (2016) the study explained that the shift of mutual assistance value is due to the business that makes people slowly start to leave the mutual assistance culture. The reason became commonplace among residents because of the lack of free time to participate in mutual assistance activities. The shift has been going on for a long time and the changes are gradually being felt until now. It is in line with the study that was conducted by Febriani (2019) the shift of mutual assistance value is caused by the level of community activity, where the community has a heterogeneous activity which causes many people to feel tired and prefer to take a rest rather than taking a part in community service.

The efforts made to overcome the obstacles faced in implementing mutual assistance values in the religious day in Balun Village were conducted by the following two attempts: (1) determining the effective time by considering the effectiveness in determining the schedule for the implementation of activities related to the mutual assistance in celebrating religious holidays; (2) giving an announcement or socialization related to the activities to be carried out, it is in line with the research result by Nilnawati and Haris (2017) that the efforts made in maintaining the mutual assistance value by providing appeals and motivating the society to maintain and preserve the mutual assistance culture is a hereditary culture and it has existed for a long time, therefore it deserves to be upheld. Besides making an event schedule, announcement or socialization is needed as an effort to make the society aware of the agenda to be carried out. It is hoped that the announcement announces in advance make the increasing of society awareness to spend their time and participate in mutual assistance activities on religious holidays. These findings can be juxtaposed with Vander's opinion

(1979) socialization is an interactive social process that affects an individual's how to think, feel, and behave so that they can participate in community life.

4 CONCLUSIONS

The mutual assistance values in religious holidays celebration in Balun Village was manifested by the existence of various activities in helping or cooperating which was carried out by the society through (1) community service; (2) donating; (3) helping the security of celebration; to (4) creating *ogoh-ogoh* statue. The process of implementing mutual assistance values still had obstacles that were faced, it was the community's activities, so not all residents could participate or help the implementation of mutual assistance. It was because when the activity took place, some residents were at work, school, and college, and also the schedule coincided with other activities. The efforts made to overcome these obstacles were by determining the effective time for the implementation of activities related to mutual assistance and providing announcements or socialization related to the activities to be carried out. Those efforts were expected could make the residents prepared themselves and took the time to participate in mutual assistance activities on religious holidays.

REFERENCES

Department of Education and Culture. 1982. *Village community mutual assistance*. Jakarta: Ministry of Education and Culture.

Creswell, J. 2012. *Research design; Quantitative and mixed qualitative approaches*. Yogyakarta: Student Library.

Febriani, et al. 2019. Factors that cause shifting values of mutual cooperation: The level of community activity and the activeness of their leaders. *Scientific Journal of Student Reasoning and Research* 3(2): 41–50.

Isbandi, R.A. 2007. *Participatory planning based on community assets*. Depok: FISIP IU Press.

Lubis, R. 2015. *Sociology of religion: Understanding the development of religion in social interaction*. Jakarta: Prenadamedia Group.

Moleong, L.J. 2007. *Qualitative research methods*. Bandung: PT Remaja Rosdakarya.

Nilnawati, R. & Haris, H. 2017. The shifting value of mutual cooperation in Tamasaju village, Galesong Utara District, Takalar Regency. *Tomalebbi: Journal of Thought, Legal Research, Pancasila Education and Citizenship* 4(4):172–187.

Nafisah. A & Sarmini. 2020. Cultural transformation of mutual cooperation in the era of globalization in the Bawean island community. *Moral and Citizenship Studies* 8(2): 383–400.

Rahmayanti. L. & Yani, M.T. 2017. Practices of religious pluralism in Balun village, Turi district, Lamongan regency. *Moral and Citizenship Studies* 5(1): 16–30.

Rismayanto, I. 2016. *The shifting of mutual cooperation values in the community of Gegerkalong village, Sukasari district, Bandung regency*. Retrieved from: http://repository.upi.edu.

Rustanto, B. 2016. *Indonesian multicultural society*. Bandung: PT. Rosdakarya youth.

Ulfatin, N. 2017. *Qualitative research methods in the field of education: Theory and its applications*. Malang: Media Nusa Creative.

Wadu L.B. et al. 2019. Mutual fundraising in Manggarai culture. *Prosiding: National Seminar on Education and Learning for Teachers and Lecturers* 3: 83–89.

Widiyanti & Sumindhia. 2003. *Indonesian Cooperatives and Cooperatives*. Jakarta: PT. Rineka Cipta.

Vander, Z.J.W. 1979. *Sociology*. New York: Jhon Wiley and Sons.

Empowering Civil Society in the Industrial Revolution 4.0 – Sukowiyono et al (Eds)

Development of local democracy in Indonesia

N. Hady* & P. Pudjantoro
Universitas Negeri Malang, Malang, Indonesia

ABSTRACT: Three important issues had been examined, namely: (1) voters behave in the implementation of the direct election of regional head general (Pilkada); (2) what was the dominant factor for voters, when they exercised their voting rights; (3) what was the idea of designing a local democracy in the future following the building construction of the unitary state of the Republic of Indonesia. This research used the mixed method with a case approach. Research results showed voter behavior was influenced by at least four things, namely: (1) aspects of voter education level; (2) the economic aspects of voters; (3) voter affiliation to certain social community organizations; and (4) the character of the prospective leader, not on the vision and mission. A direct democratic design was required that places it at the provincial level, while for Pilkada in a regency or city, it is sufficient to elect through the Regional House of Representatives (DPRD).

Keywords: development, local democracy, Indonesia

1 INTRODUCTION

Local democracy in Indonesia has experienced development, especially after the reformation, where previously regional heads, both Governors, Regents and Mayors were elected by the local parliament, but post-reform has experienced developments, namely being directly elected by the people in the regions, although Article 18 paragraph (4) of the 1945 Constitution does not mandate direct regional head elections, it only affirms that they are elected democratically, but the political law then regulates direct elections. It is in this context that the authors see that there has been a shift from representative democracy which upholds the principle of deliberation and consensus, to direct democracy.

Of course, the shift from representative democracy to direct democracy in regional head elections has implications for several things, both from the aspects of political participation, political culture, voter behavior, and too high political costs. This research attempts to link the implications of shifting the implementation of local democracy in Indonesia from these various aspects.

2 METHODS

This study uses a combined research approach according to Hesse-Biber (2010), which includes the collection, analysis, and integration of quantitative and qualitative data in a single or phased study (Jonathan 2011). The quantitative data is based on data from the field survey results of the regional head elections in Malang City and Malang Regency, with a total of 130 respondents, while the qualitative data is based on media analysis data related to the implementation of regional elections in Indonesia.

*Corresponding Author

DOI 10.1201/9781003180128-16

3 RESULTS AND DISCUSSION

The implementation of local democracy in Indonesia is experiencing interesting dynamics to study. In the New Order era, local democracy was largely determined by the central elite because the main determinant of victory was in the blessing of President Soeharto through the Ministry of Home Affairs, so that the position of local parliament was only to propose candidates for a regional head, the rest was the central government so that at that time the dominant one was elected is a candidate from a military figure, especially when ABRI's dual doctrine is still very strong. Marijan (2011) called this because of the centralization of power so that in such a centralized political structure, the forces of society play a more role as the audience of a large political theater that does not have the power to influence the policies that have been decided. The following developments, entering into the reform era that shifted the pendulum of power in the context of regional head elections were entirely in the local parliament, the central government was limited to approving it. It's just that even at this time the nuances of back and forth are still so pronounced, there was even conflict with local political elites in several regional elections (Mashad 2005), as well as massive money at the local parliament level, because the one who determines the election of a regional head is a member of the local parliament, so at that time the local parliament is very strong and strategic.

Also, the relationship between the regional head and the DPRD is less harmonious, because the local parliament can dismiss the regional head at any time if the regional head is unable to build good relations with the local parliament. That was one of the reasons for the emergence of the idea of direct election. Direct regional head elections are expected to open more doors for the emergence of regional heads by the wishes of the majority of the people, in addition to maintaining government stability so that it is not easily overthrown in the middle of the road (Hamidi, et al. 2010), as well as being the first step for strengthening the role community (Zuhro 2011). Also, direct regional head elections can accommodate an integrated selection system that complements each other to produce qualified candidates for elected regional heads, so that they become an integral part of the acceleration of democratization at the national level (Mahfud 2012), and is also expected to reduce widely the existence of what Marijan (2011) called "power hijacking" carried out by political parties that have seats in the local parliament, as well as being able to produce regional heads have higher accountability to the people. Even so, departing from the process of direct regional head elections which began on June 1, 2005, we have not seen that this hope has not yet fully come true. This is because, with the direct Pilkada, we are still unable to produce quality regional leaders who are committed to building their regions properly and free from corruption.

The direct Pilkada process in the view of Mahfud (2012), was still colored by various frauds, such as money politics with various modes, the rise of moral pragmatism, both for candidates and the community, perpetuating the oligarchy of power, triggering bureaucratic politicization, becoming an arena for power rivalry. in an unhealthy manner, so as not to produce leaders with political virtues who act responsibly, prioritize the interests of the community above personal, group, or party interests, and the Pilkada is directly vulnerable to political elite conflicts involving the masses. From the aspect of political participation, the results of research in several direct regional elections in Indonesia, the level of participation is still very low, which is below 80 percent, for example, the Malang City Pilkada in 2018 was only 65%, the local elections in 2020 in Pasuruan City were 76.3%, even The direct regional election for Surabaya was only 52.4%, and the Pilkada in Malang Regency had only 58% voter turnout.

The low level of public voter participation is caused by many factors, one of which is that the community is already saturated with the direct Pilkada, especially the implementation of the Pilkada during the Covid-19 Pandemic which of course raises concerns from the aspects of public safety and health. Also, this is due to the perception of voters who say that the Pilkada is useless for improving the situation, namely 30.8%, 17.7% who think there is no favorite candidate pair and 12.3% think the Pilkada benefits the Candidate only.

From the aspect of political culture that related to the views and attitudes of individuals in society as fellow citizens (Almond & Powell 1978), society as voters is immature in responding to political differences, even fellow campaign teams attack each other with words that already lead to a black campaign. Not to mention the problem of political conflicts that often occur, especially in the early days of direct Pilkada implementation, although in the view of Almond & Verba (1990), that cooperation and conflict between groups or social groups is an actual feature that can color political culture in society, of course, if this reality occurs in direct Pilkada events it will seriously injure democracy, so that the direct Pilkada which is a moment of democratic party shifts into a political phenomenon which is still marked by problems, such as the rise of political dynasties, massive money politics with various modes, and political conflicts.

From the aspect of voter behavior, which Surbakti (1992) defines as the participation of citizens in elections to exercise their voting rights which are influenced by five factors, namely: structural, sociological, ecological, social-psychological, and rational choices. The results of research on Pilkada in Malang Regency in 2020 provide information that from the aspect of the structural approach, the majority of voters, namely 90.2%, know that there will be Pilkada through various sources, both through political parties, community leaders, mass organizations and information from election organizers, as well as information on Pilkada in Malang City as much as 94.00% know the information about the Pilkada. In the sociological approach, most voters, namely 56.2%, did not see the choice based on the gender of the candidate.

Also, 29.2% of voters who used their voting rights said it was a citizen's right and 29.2% said they joined other people, followed by 26.9% who wanted Malang Regency to be even better, and 11, 5% want to have an even better Bupati. The voters' perception above is very reasonable because the Malang Regent was previously caught in a corruption case by the Corruption Eradication Commission (KPK) and is still in legal proceedings, so of course, voters do not want it to happen again.

Meanwhile, their reasons for choosing a leader were based on the reason for being able to overcome economic problems, namely 22.3%, then 13.8 because of the character of the candidate, the rest because the candidate was honest, clean, and socially spirited. Voters' perceptions that are not attractive are the main considerations in making choices because the candidates can solve problems in Malang Regency which is 73.1%, followed by 10.8% from the personality of the Regent candidate. From the aspect of the candidate's personality, most of the voters, namely 51.5%, were honest, followed by 15.4% who were popular candidates or close to the people, followed by 14.6% who were able to solve problems. Another interesting voter behavior is that voter choice is based on ethnic similarity, namely 45.5%, followed by 40% based on religious similarity, the rest is based on similarity in parties, social status, equality in profession, gender equality, and similarity in social community organizations, the percentage is below 5%.

The ecological approach is very visible when voters who are in the West and East Malang areas whose access to public services is still difficult to reach, where the majority of the population work as farmers. Therefore, the perception of voters wanting candidates to be able to solve economic problems was 34.6%, followed by 20.8% to be able to solve social problems and be able to solve corruption problems by 13.1%, and the rest were able to solve bureaucratic problems, environmental damage, and infrastructure problems.

From the aspect of political costs which are certainly too high, both the costs for Pilkada organizers and the costs that must be incurred by candidates for the regional head. The budget for the 2020 Pilkada for Surabaya City is 101.24 billion, the Pilkada budget for Malang Regency is 85 billion, and the Pilkada for Pasuruan City is 23 billion, not to mention the budget spent by the candidates, both for the cost of printing t-shirts, banners, banners, and the costs for limited meetings. The budget issued was so large that it led to cases of corruption that ensnared several regional heads in Indonesia.

4 CONCLUSIONS

When democracy has not yet provided prosperity, it will turn into doubt and even despair. Even though local democracy in Indonesia is experiencing fast dynamics after the direct Pilkada, but of course it is also necessary to evaluate the substance of its implementation, whether it has brought benefit to the people, or vice versa. Within that framework, the development of local democracy is not determined solely because it is directly elected, but what is most important is the extent to which the Pilkada can build a democratic and clean local government and bring prosperity to the people in the regions, if that reality has not been realized, then, of course, we need to rethink Ir. Soekarno's opinion, that: "...if we are looking for democracy, it should not be Western democracy, deliberation that gives life, namely political economy, democracy capable of bringing social welfare".

REFERENCES

Almond, G. & Powell, G.B.P.J. 1978. *Comparative politics*. Boston dan Toronto.

Almond, G. & Verba, S. 1990. *Budaya politik tingkah laku politik dan demokrasi di lima negara* (S. Simamora, Trans). Jakarta: Bumi. Aksara.

Hamidi, J. et.al. 2010 *Rethinking penyelenggaraan pilkada yang demokratis dan partisipatif dalam konstitusionalisme demokrasi.* Malang: Penerbit Intrans Publishing.

Hesse-Biber, S.N. 2010. *Mixed methods research: Merging theory and practice*. New York: Guilford Press.

Jonathan, S. 2011. *Mixed methods.* Jakarta: Penerbit Elex Media.

Mahfud, M. 2012. Evaluasi pemilukada dalam perspektif demokrasi dan hukum. *In Demokrasi Lokal, Evaluasi Pemilukada di Indonesia*. Jakarta: Konstitusi Press.

Marijan, K. 2011. *Sistem politik Indonesia.* Jakarta: Penerbit Kencana Prenada Media Group.

Masdar, H. 2017. *Jalan demokrasi kita.* Malang: Penerbit Intrans Publishing.

Surbakti, R. 1992. *Memahami Ilmu Politik.* Jakarta: Penerbit Grasindo.

Zuhro, R.S. 2011. *Model demokrasi lokal, Jawa Timur, Sumatera Barat, Sulawesi Selatan dan Bali.* Jakarta: Penerbit THC Mandiri.

Empowering Civil Society in the Industrial Revolution 4.0 – Sukowiyono et al (Eds)

Constructing the student political commitment based on Indonesia's national identity

R.A. Mawarti*, S.A. Hakim, S. Untari & M.M. Habibi
Universitas Negeri Malang, Malang, Indonesia

ABSTRACT: This article discussed the formulating process of students' political commitment based on Indonesia's National Identity. The method used a qualitative approach with case study design on Law and Citizenship department student, Faculty of Social Sciences, Universitas Negeri Malang w, studying political education. Researchers collected through the stages of preliminary study, observation, interviews, and documentation study. The results showed that a learning process executed with deliberation and discussion could enhance students' understanding of good citizen roles and commitment to Indonesia's national identity. Students were able to express their thoughts in various art-products of political commitment. The students have expressed the commitment through poetry readings, drama performances, song presentations, and other artistic expressions. So that indirectly, it also creates a sense of nationalism following Pancasila, not chauvinism.

Keywords: political commitment, Indonesia's national identity, good citizen

1 INTRODUCTION

Nowadays, the construct of The Unitary State of the Republic of Indonesia is still the most frequently questioned among Indonesian, and it is all about the possibility that it will change or it will be the final answer. Despite the amendment of the Constitution of the Indonesia Republics (UUD NRI 1945), Article 37 Paragraph (5) has affirmed that "specifically the form of the state cannot be amended," but still so vulnerable to the nation's commitment to establishing the existence of Indonesia as a unitary state (Al-Hakim 2014).

On the other side, globalization is not a new phenomenon, but today, the internet is one of globalization's most visible facets (Kirchner 2009). This part of the era spread the 'virus' of global perspective and held up energy nationalism. Optimistic assumptions can see energy nationalism about the domestic energy projects perceiving them as safe and harmful assumptions about the foreign projects, perceiving them as unsafe (Cesnakas 2013). Globalization paradoxical also offers many global perspectives that can give a negative-positive impact, and this nation may need to do a national reflection, especially in building a cultural-ideological movement to strengthen the national identity and existence.

More specifically, the main problem that is faced by Indonesian people in the Global era nowadays is still much poverty, low human resources, and social inequality. Rodrigo A. Chaves (The World Bank 2016) said that "Indonesia is at risk of leaving its poor and vulnerable behind. Poverty reduction has begun to stagnate, with a near-zero decline in 2014. Income inequality is rapidly rising, and up to one-third of it is explained by inequality of opportunities. Healthy and well-educated children live side by side with children who suffer from malnutrition, learn little when they are in school, and drop out too early".

*Corresponding Author

DOI 10.1201/9781003180128-17

From that issues, if the Indonesian people could explore the national lives again and then implement it, then globalization will be dealt with efficiently, all of the Indonesian people could solve the problems, and the Republic's integrity will remain intact. In this context, both national knots and pillars need to revitalize, especially its culture strategy as a 'guardian' of the Unitary State of the Republic of Indonesia in the middle of the gloomy shadow of 'global ideology.' A variety of global ideologies that can threaten Indonesia's integrity can be faced with promoting a wise, prudent, moral, and religious attitude not to lose the real definition of The Republic of Indonesia's Unitary State.

This commitment becomes firmly embedded in Indonesia's national or cultural identity. Cultural identity is necessarily bound to a social group and a specific place, such as a nation-state (e.g., Ukraine) or a particular region within a nation-state (John-Stewart 2016). The character of children and students of Indonesia. The purpose is to make them know better and more love all about Indonesia. They know about it; this will help decide the Indonesian ability to manage their homeland. Through that effort, wishing entirely, they always had a kink and high commitment to building a nation based on the family spirit of unity and Indonesian multiculturalism in the Unitary State of the Indonesian Republic context.

Meanwhile, political and student terminology is still often debated. Some opinion said that student did not interest in political topics. Patterson, T. noted, "Today's young adults are less politically interested and informed than any cohort of young people on record" (Patterson 2002). The other opinion says that students do not need to thinking of politics. Politics is only for adults, and they are still too young to get involved in that topic. Student involvement in the political sphere (read: political) practically will cut academic freedom.

Based on that role, there is should no more distance between students and political life. The campus should use a medium that embodies the student to learn about the community's political life. As Putnam puts it, "If you do not know the rules of the game and the players and do not care about the outcome, you are unlikely to try playing yourself" (Putnam 2001). In other words, without learning politics, a student cannot raise as a good citizen. Because fundamentally, citizen life is related to politics. The existing efforts can start from constructing political learning in the classroom, electing the campus organization's chairman, and engaging them in political process and practices. They have their commitments according to their organizations' political labels and how they seek to advance their organizations.

In the research contexts, student political commitment based on Indonesia's national identity has no big count. Another researcher was more interested in talking about government political commitment or another subject. So, we decide to take care of students as the main subject of political commitment and use Indonesia's national identity as the object. Because the main result of Indonesia education programs, especially citizenship education, is raising a good citizen who could implement national identity values.

Then, several things become the goal of this research: (1) Student mechanism in building political commitment based on Indonesia's national identity; (2) Various forms of products of students' political commitment to the Republic of Indonesia; (3) Positive impact felt by students from political commitment based on Indonesia's national identity. The existing research is expected to new references about the learning process to encourage students to become good citizens, commit, and love the homeland.

2 METHODS

The study used a qualitative approach and case study research design to Law and Citizenship department student, Faculty of Social Sciences, Universitas Negeri Malang. The participants or the main subject of research were the students taking political education courses and four classes—the study located in the Law and Citizenship Department, Faculty of Social Sciences, Universitas Negeri Malang.

Data collection techniques were qualitative observation, qualitative interviews, Focus Group Discussion, and qualitative documents. Data analysis used the theme incorporation in 3 phases: reduced the data that corresponds to the study's focus, presented the class of findings, and found out the conclusions. For the data validity, the researchers used a technique of triangulation. In this regard, the triangulation technique used source triangulation and theory triangulation, where the data of political commitment was collected and then clarified with data sources (students) and with the theory/opinion (expert). After obtaining the validity, the researcher analyzes it based on the analysis technique used in this research.

3 RESULTS AND DISCUSSION

3.1 *Mechanism of students in formulating political commitment*

In the existing research, political commitment formulation was taken in several ways, broadly developing deliberation. It is evident in some groups who claim that they explicitly use deliberation with FGD way, but some groups claim that various discussion is theirs. In the group of who uses focus group discussion, the following stages of the political commitment formulation are: (a) focusing the attention; (b) focusing the discussion on the context of the "political education platform"; (c) identifying the political, educational foundation, sort and find which later become formulated within the political commitment; (d) Discussing and identifying together, sharing and mutual express opinions to achieve a conclusion that will be approved by all parties; (e) writing the political commitment formulate. For example, the first group who get the theme of commitment to the August 17th, 1945 proclamation in a structured running the stages FGD with the main focus of committing to the event.

Based on the findings, the researchers analyzed the nature and benefits of the discussion deliberation method, especially FGD. Though there exist multiple understandings of public deliberation, all agree that one valuable deliberative practice is conducting a substantive discussion among a small number of persons in face-to-face settings. In such discussions, participants can develop a stronger sense of civic identity and self-efficacy, augment their political knowledge and communication skills, and refine their political judgments (Gastil 2006). Furthermore, Champions of public deliberative forums maintain that the experience of making, hearing, and evaluating arguments makes people more politically knowledgeable as well as more reasonable (Francesca et al. 2009). In other words, we know that the method of deliberation is full of benefits in the effort of discussion to solve the problem.

More specifically, one deliberation type is Focus Group Discussion. A focus group discussion is an in-depth, open-ended, qualitative group discussion, lasting one to two hours, that is employed to collect information from a few individuals to provide data on a predefined topic (Akpabio et al. 2007). This definition is related to the method of deliberation. Both are generally beneficial to anyone who uses them seriously and complies with existing procedures. In college learning, discussion methods with FGD types help students solve problems and understand materials. Thus, they can exchange information and learn to respect each other's differences of opinion and make the best decisions for all group members.

Researchers can conclude that the students' mechanisms in formulating political commitment to the Unitary Republic of Indonesia using the method of discussion deliberation and FGD are the right way. Existence is the appropriateness between deliberation with the experts' thinking and the accuracy of the benefits obtained. With Focus Group Discussion to seek the common good, each member in the group gets an opportunity to propose what is best for the group. In the context of the state, this kind of thing is in line with the implementation of Pancasila democracy, especially the fourth precept. In other words, students also have learned to perform their roles or obligations as good citizens.

3.2 *Political commitment product of students to Indonesia's national identity*

Through the process of formulating the political commitment based on Indonesia's national identity, which taken by students, found some products as follows in table 1. Various kinds of discussion results presented by the group could be regarded as a political commitment product based on several considerations. First, the resulting product tends to contain a promise or pledge of allegiance to something in line with the concept developed by Robbins (1993). Secondly, the content of political commitment appears to be the expression of love or loyalty aimed at nationality and statehood values, especially in Indonesia's national identity content. It is in line with the notion of political commitment from Park (Zurnali 2010).

Table 1. Poem and yell with the commitment object.

Group 1 Commitment object: Garuda Pancasila Product: Poem Title: My Garuda Pancasila	Group 2 Commitment object: Youth Pledge Product: Yell Title: Invincible youth
I know mighty, my Garuda I know the firmness in your chest On the stake, I hope The world will always be vigilant, Will the robustness of my Garuda Pancasila? Rise my garuda, flutter your wings Flying high over the clouds Prepare your sharp claws Shreds of their armor Until no one can protect evil In my beloved land, Indonesia	We are the Indonesian youth generation We support the unity of the nation Pancasila as the primary, UUD NRI 1945 as the source of law, and Youth Pledge as the unity value "United youth cannot be defeated."

3.3 *The group perceives the benefit of formulating political commitment*

Political education implementation packed with activities to formulate a political commitment by the students of Indonesia's national identity has shown a positive impact. The first positive impact generated is increasing students' understanding and urgency sense of the political commitment. Students actively engage in dialogue or discussions during the learning process that encourages them to cooperate and think critically. Besides, through improved understanding is, and then evolved several other positive impacts, among others:

a. Fostering a sense of nationalism of Indonesia's national identity;
   It is clear when student groups are diligently formulating a political commitment based on Indonesia's national identity as outlined in commitment products. Where all the products produced contain national values. Thus, this becomes one of the proofs of the students' love of Indonesia's national identity and the values that are in it.
b. Strengthening relationships between members of a group or fostering solidarity;
   This impact is one of the most visible impacts of the eye. The students work together in teams that require every member of the group's contributions. Without their cooperation and solidarity, the group will not make a strong political commitment based on Indonesia's national identity.
c. Training members of the group to be brave and responsible in making political commitments;
   This impact is one of the effects that may not become sufficiently visible but becomes a substantial impact to know. When students are encouraged and established to formulate the political commitment to the homeland and run optimally, indirectly, they have shown courage and a responsible attitude.

Courage appears when they dare to formulate something that then binds their words, thoughts, and actions in the future as citizens. At the same time, responsibility is evident when they proclaim existing commitments. With these activities, students will learn to run things according to what he had promised or committed.

4 CONCLUSIONS

Student political commitment based on Indonesia's national identity is an important learning activity to do. It is based on the idea to face globalization development that merges the boundaries between countries and facilitates the exchange of values; students who are semi-finished citizens should be given a strong foundation of a commitment to nationality values. The learning process can be constructed on the learning that gives the students an active role in finding the source of learning and achieving the learning objectives. One of the methods used is through deliberation method, especially FGD or varied discussion type. With supervising and coaching, students formulate commitments to Indonesia's national identity with various commitment expressions that deserve to be appreciated. This process is the one that then affects the growth of a sense of nationalism, solidarity, and courage, and responsibility. Problems that arise in the form are lack of communication running smoothly and so forth must be handled not to be an obstacle that stops achieving goals.

REFERENCES

Akpabio, et al. 2007. Utilisation of focus group discussion (fgd) as a research tool in community health nursing practice: A case study of the views and beliefs of secondary school students in Nigeria about HIV/AIDS and its prevention. *Africa Journal of Nursing and Midwifery* 9(1): 78–90.

Al-Hakim, S. 2014. Pendidikan kewarganegaraan dalam konteks Indonesia. Malang: Intrans Publishing.

Cesnakas, G. 2013. Energy security challenges, concepts and the controversy of energy nationalism in lithuanian energy politics. *Baltic Journal of Law & Politics* 6(1): 106–139.

Francesca, et al. P. 2009. Is information good for deliberation? *Journal of Public Deliberation,* 5(1): 1–20.

Gastil, J. 2006. How balanced discussion shapes knowledge, public perceptions, and attitudes: A Case study of deliberation on the los alamos national laboratory. *Journal of Public Deliberation* 2(1): 1–32.

John-Stewart, G. 2016. Human rights and cultural identity. *Baltic Journal of Law & Politics* 8(2): 112–135.

Patterson, T. E. 2002. *The vanishing voter: Public involvement in the age of uncertainty.* Cambridge, MA: Alfred A: Knopf Publishers.

Putnam, R.D. 2001. Bowling alone: *The collapse and revival of American community.* Simon and Schuster.

Robbins, S.P. 1993. *Organizational behavior concepts, controversies, and applications.* Englewood Cliffs: NJ: Prentice-Hall.

The World Bank. 2016. Indonesia's Rising Divide. Jakarta: The World Bank Office.

Zurnali. 2010. *Learning organization, competency, organizational commitment, dan customer orientation: knowledge worker – kerangka riset managemen sumberdaya manusia di masa depan.* Bandung: UNPAD Press.

Empowering Civil Society in the Industrial Revolution 4.0 – Sukowiyono et al (Eds)

The problems of developing a tourist village

M.I. Alawi*, M.M. Habibi & N.W. Rochmadi
Universitas Negeri Malang, Malang, Indonesia

B.S. Wiwoho
University of Queensland, Brisbane, Australia

ABSTRACT: This study aimed to explain the role of village government in tourism development and identifying the inhibiting factors in the development of the Sendang Bulus Beji Pager Tourism and solutions to these obstacles. This research used a qualitative approach with a descriptive research type. The data collection techniques used were in-depth interviews, observation, and documentation. The results showed that the role of the village government in tourism development was to initiate the development of the Sendang Bulus Beji Pager Tourism. As the supervisor and person in charge of the development of the Sendang Bulus Beji Pager Tour and set the Sendang Bulus Beji Pager Tourism to be a Village-Owned Enterprise. The inhibiting factors for tourism development are limited funds, the absence of good tourism administration management. An effort to overcome the existing obstacles is that the Pager Village Government submits a request for assistance to the central government in the form of channeling funds and optimizing the use of village funds provided for tourism development and the updating of tourist rides and facilities.

Keywords: village government, Sendang Bulus, tourism

1 INTRODUCTION

Tourism has an important role in the economic development of a region. The economy of an area can increases if tourism in the area develops because many tourists come. Tourism activities can encourage and accelerate economic growth. In line with this, the impact of tourism on the socio-economic conditions of local communities is grouped by Cohen (in Pitana & Diarta, 2009) into eight major groups, namely: (1) impact on foreign exchange earnings, (2) impact on community income, (3) impact on employment, (4) impact on prices, (5) impact on distribution of society or profits, (6) impact on ownership and control, (7) impact on development in general and (8) impact on government revenues.

Regions in Indonesia have many potential tourist destinations that are scattered throughout the archipelago. Ponorogo Regency is a district located in East Java Province which has high potential resources that can be empowered by its tourism sector. There are several tourism potentials in Ponorogo Regency, including nature tourism, religious tourism and cultural tourism. Quoted from (Tempatwisataseru.com) in Ponorogo Regency, there are around 46 tourist destinations ranging from historical tourist attractions, natural attractions, museums, and water games that can be relied on for development.

Pager village is a village located at the eastern end of Bungkal District, 30 kilometers from downtown Ponorogo. There is a natural tourist attraction which is the only one in Ponorogo Regency and not found in other districts, namely Sendang Bulus. Sendang which means pond and bulus means softshell turtle. Since a long time ago, this spring has been known as the original habitat for

*Corresponding Author

 DOI 10.1201/9781003180128-18

bulus. This spring has become an attractive tourist area because it offers a beautiful and unspoiled panorama, but because of the large amount of hunting and turmoil that has occured near the spring, this natural tourism has been neglected for years and has been deserted by visitors. So far, this spring is only used as a reservoir for irrigation for village farmers who own rice fields near the Sendang Bulus Beji Pager Tour. In 2015 this place began to be developed again until now it has become one of the famous tourist locations in Ponorogo Regency.

Sendang Bulus Beji Pager Tourism has several tourism potentials. The most important thing for the tourism potential is the existence of a spring or spring whose water is always available. The abundant water is usually used to irrigate rice fields. The next tourism potential is that this place is still very natural, the trees are also still shady plus the south and west of Sendang there are expanses of rice fields and mountains. Furthermore, there is the maintenance of freshwater fish, including tawes, tilapia, tombro fish and snakehead fish and fleas. Visitors can also feed directly to the fish that are kept there. The manager has provided feed in the form of pellets which are priced at 2000 per pack. In addition, visitors can also enjoy the view by riding a duck boat.

The role of the Pager Village government and all components in the community is very necessary, considering the potential in Sendang Bulus Beji Pager is very good to be developed. The development of a tourist attraction requires involved stakeholders, human resources, programs, funds and facilities. Based on the explanation above, the researcher is interested in researching "The Problems of Developing a Tourist Village." The researchers' reason is interested in taking the title is to find out what the role of the village government is in the development of tourism, where previously the tourist spot was not maintained and not managed until now it is developed and better managed so that now Sendang Bulus Beji Pager is a famous tourist destination in Ponorogo Regency.

2 METHODS

The location of this research is in Pager Village, Bungkal District, Ponorogo Regency. This study aims to explain the role of village government in tourism development. Identify the inhibiting factors in the development of the Sendang Bulus Beji Pager Tourism and seek solutions to these obstacles. This research uses a qualitative approach with descriptive research type. There are three sources of data in this study, namely humans, documents, and events. The data collection techniques used were in-depth interviews, observation and documentation. Researchers conducted interviews with several informants from Pager Village, researchers used observation to observe directly and record the conditions in the tourist environment of Sendang Bulus Beji Pager Village and Pager Village itself. The documents used in this study were documents obtained from informants regarding the role of the village government in developing the tourism development of Sendang Bulus Beji Pager in Pager Village, as well as photos during field observations and interviews with several informants. The data analysis technique used was the Miles and Huberman data analysis technique, namely data collection, data reduction, data presentation and conclusion drawing.

3 RESULTS AND DISCUSSION

Pager Village is located in Bungkal District, Ponorogo Regency, East Java Province, Indonesia. Pager Village consists of 3 hamlets, namely Bibis hamlet, Glagah Malang hamlet, and Pager Tengah hamlet, each of which is led by a head of the hamlet. In order to maximize the function of service to the community in Pager Village, the three hamlets are divided into 6 *Rukun Warga* (RW) and 18 *Rukun Tetangga* (RT). The pager village has an area of about 460 Ha with a population of 1,847 people with 632 the number of heads of households. The role of village government in tourism development of Sendang Bulus Beji Pager.

Tourism has an important role in the economic development of a region. The economy of an area can increase if tourism in the area develops because of the many tourists who arrive. As one of the potential aspects in increasing regional and community income, tourism must be well developed. To develop a tourist object or place, it is necessary to have a role from the government and the community in its management. Tourism management must be carried out in a planned and comprehensive manner. So that optimal benefits can be obtained for the community, both from an economic, social and cultural perspective.

The Village Government of Pager is the party responsible for and is the supervisory party in the development of Sendang Bulus Beji Pager Tourism. The village government provides the opportunity for tourism conscious groups and the community to use these heritage springs to be maximally empowered so that the place can be even more useful for all villagers. One of the things that can be done to take advantage of the potential of this place is to build and develop the Sendang Bulus Beji Pager Tourism. According to Law Number 6 of 2014 Article 76 concerning Village spring ancestral heritage is included in the category of village assets. Then in article 77 paragraph 2 it is said that the management of village property is carried out to improve the welfare and standard of living of the village community and increase village income. In addition, the role played by the Pager Village government is in line with Law Number 6 of 2014 article 26 paragraph 2 c which states that in carrying out its duties the Village Head has the authority as the holder of financial power and village assets. So, the village government in this case has the authority to grant permission to tourism conscious groups and community members to manage and utilize the assets of the Pager Village.

The development of Sendang Bulus Tourism can be categorized as village development. This is based on Law Number 6 of 2014 article 78 concerning Villages which states that village development aims to improve the welfare of rural communities and the quality of human life as well as poverty alleviation through meeting basic needs, building village facilities and infrastructure, developing local economic potential, and sustainable use of natural resources and the environment. The use of Sendang Bulus Beji Pager as a natural tourism has the potential to empower the community in a sustainable manner. This role was continued by including the Sendang Bulus Beji Pager Tourism into a village-owned enterprise called Sendang Wiro Prenggo through Village Regulation Number 04 of 2016 concerning the Establishment of Pager Village Owned Enterprises, Bungkal District, Ponorogo Regency on April 15, 2016. It is hoped that after its entry Sendang Bulus Beji Pager Tourism into a village-owned enterprise so that its development can be more planned and its development can be taken from village funds.

3.1 *Inhibiting factors in the development of Sendang Bulus Beji Pager tourism*

Tourism Development of Sendang Bulus Beji Pager is inseparable from constraints both from within and from outside. The main inhibiting factor for the development of the Sendang Bulus Beji Pager Tourism is the lack of funds for tourism development. Although the Sendang Bulus Beji Pager Tourism has entered a village-owned enterprise Sendang Tirto Prenggo belonging to Pager Village, in its development, it cannot be entirely taken from village funds. So that in the development of these tours requires a relatively long time and must be done gradually. The long and long process in developing this tourism makes visitors even more quiet, only on Saturdays and Sundays and only on holidays which are crowded with visitors considering that the Beji Pager Bulus Tourism is still under construction, making visitors who come uncertain. Suwantoro (2004) explains that one of the main elements that must be considered in supporting tourism development is one of the objects and tourist attractions. When the attractiveness of a tour weakens, the visitors who will come will also gradually decrease.

The next inhibiting factor is the lack of good tourism administration management. Filing that is not clear regarding the recapitulation of the number of visitors who come every day or about finances is still very simple. In addition, the Sendang Bulus Tourism, which is still under construction and has not been fully realized, as designed in the master plan is the reason for the temporary not being subject to tourist entry fees. What is feared by the manager if the tourism condition is still as it is now, the fee will result in a decrease in the number of visitors. This is still not in accordance with Law Number 6 of 2014 article 78 paragraph 2 concerning Villages that village development includes the planning, implementation and supervision stages. This article indicates that the development of a village must go through a planning stage that has a good time frame and structure. Even though all the plans for the development of the Sendang Bulus Beji Pager Tourism have been neatly arranged and have been documented in the form of a master plan, the administrative management of the tour is still not good.

3.2 *Efforts made to overcome obstacles in the development of the Sendang Bulus Beji Pager tourism*

The village government together with the management and the community must work together in overcoming the obstacles that become obstacles in the development of the Sendang Bulus Beji Pager Tourism. The most important obstacle is related to the lack of funds to develop this tourism. What the village government does is submit a request for financial assistance to the central government in the form of funds *sharing*. In addition, the village government has also submitted requests for assistance to the provincial government through the House of People's Representative aspiration program. Submission of requests for assistance to the central and provincial governments has been made even though they have not been received because all promises have not been realized. In addition, the village government must communicate frequently and often coordinate with all related parties and continue to establish relationships with the community to make investments or community investments.

To overcome obstacles related to village government funds and tourism managers, they try to optimize the use of village funds provided for tourism development. Development is carried out slowly and gradually every year so that the funds used can be optimized. The use of village funds in the process of developing the Sendang Bulus Beji Pager Tourism in accordance with Law Number 6 of 2014 article 74 paragraph 1 regarding Villages explains that village expenditure is prioritized to meet development needs agreed upon in village meetings and in accordance with the priorities of the regional/city government, government provincial and government areas.

The development process that continues in stages as planned in the master plan, especially the completion of the construction of the entrance gate for the Sendang Bulus Beji Pager Tourism, as well as the addition of new rides using village funds is expected to increase the number of visitors who come. This is in accordance with the opinion expressed by Yoeti (2008) that one of the things that can make a tourist object interesting is that there is something to do. Something to do in this case means that the location or tourist spot is not only visible but also can support tourists to do something. The addition of playgrounds, outbound, duck boats, odong-odong, providing fish feed at a price of 2000 per pack, and the presence of a swimming pool will make tourists feel more at home in the Sendang Bulus Beji Pager Tourism because there is something to be done, namely to surround the Sendang Bulus Beji Tourism. By riding a boat, children can also play in the playground, can ride odong-odong, eat fish as much as they want, and can also swim.

If visitors who come to tourist attractions are crowded, the stalls around the tour can get income from selling food and drinks. In addition, when the visitors who come are getting more crowded and the construction of the Sendang Bulus Beji Pager Tour is quickly completed and the tourist rides are also being added as planned, especially the gate entrance as the counter door, it will be able to immediately support improvements in the administration of visitor data collection and matters relating to data collection on the tourism development program.

4 CONCLUSIONS

The village government is one of the parties that plays an important role in tourism development. The role of the village government in the development of the first Sendang Bulus Beji Pager Tourism is as the initiator of the initial development of the Sendang Bulus Beji Pager Tourism. The second role is as a supervisor as well as participating as a person in charge of developing the Sendang Bulus Beji Pager Tourism. The third role is to provide permits and mobilize a tourism awareness group with the community to develop village assets that have not been maintained for a long time. The fifth role is to make tourism a village-owned enterprise. However, there are several inhibiting factors in the development of the Sendang Bulus Beji Pager Tourism. The first factor is limited funds for tourism development. The next factor is the lack of good tourism administration management. Efforts made to overcome obstacles in the development of the Sendang Bulus Beji Pager Tourism are the Pager Village Government applying for financial assistance to the central government in the form of fund distribution and optimization of the use of village funds provided for tourism development, vehicle renewal. and adding new facilities to ensure the Sendang Bulus Beji Pager Tourism.

REFERENCES

Akbar. H.U & Purnomo, S. 2009. *Metodologi penelitian sosial*. Jakarta: Bumi Aksara.

Antariksa, B. 2016. *Kebijakan pembangunan kepariwisataan (Pengembangan kepariwisataan yang berkelanjutan dan perlindungan kekayaan intelektual).* Malang: Intrans Publishing.

Arikunto, S. 1996. *Prosedur penelitian suatu pendekatan praktek*. Jakarta: PT. Rineka Cipta.

Djafar, S. 2015. *Evaluasi kebijakan pariwisata*. Yogyakarta: Penerbit Ombak.

Martaleni. 2017. *Pemasaran pariwisata di Indonesia*. Malang: International Research and Development for Human Beings.

Maryani, E. 1991. *Pengantar geografi pariwisata*. Bandung: Jurusan Pendidikan Geografi IKIP Bandung.

Moleong, L.J. 2007. *Metode penelitian kualitatif.* Bandung: Remaja Rosdakarya Offset.

Moleong, L.J. 2014. *Metode penelitian kualitatif edisi revisi*. Bandung: Remaja Rosdakarya.

Pitana, I.G. & Diarta, I. K. S. 2009. *Pengantar ilmu pariwisata*. Yogyakarta: Penerbit Andi.

Priasukmana, S. & Mulyadin, M. 2001. Pembangunan desa wisata: Pelaksanaan Undang-Undang otonomi daerah.*Jurnal: Info Sosial Ekonomi* 2(1): 37–44.

Soedarmayanti. 2014. *Membangun & mengembangkan kebudayaan & industri pariwisata (Bunga rampai tulisan pariwisata).* Bandung: Refika Aditama.

Sugiyono. 2016. *Metode penelitian pendidikan pendekatan kuantitatif, kualitatif dan R&D*. Bandung: Alfabeta.

Sujali. 1989. *Geografi pariwisata dan kepariwisataan*. Yogyakarta: Fakultas Geografi Universitas Gadjah Mada

Sukmadinata. 2013. *Metode penelitian pendidikan*. Bandung: Remaja Rosdakarya

Sukriono, D. 2010. *Pembaharuan hukum pemerintahan desa*. Malang: Setara Press.

Suwantoro, G. 2004. *Dasar-dasar pariwisata*. Yogyakarta: Penerbit Andi.

Republik of Indonesia. 2009. Undang-Undang Nomor 10 Tahun 2009 tentang Kepariwisataan.

Republik of Indonesia. 2014. Undang-Undang Nomor 6 Tahun 2014 tentang Desa.

Wardiyanata. 2006. *Metode penelitian pariwisata*. Jogjakarta: Penerbit Andi.

Yoeti, A.O. 2002. *Perencanaan dan pengembangan pariwisata*. Jakarta: Cetakan Pertama Pradnya Paramita.

Yoeti, A.O. 2008. *Ekonomi pariwisata: introduksi informasi dan implementasi*. Jakarta: Kompas.

Empowering Civil Society in the Industrial Revolution 4.0 – Sukowiyono et al (Eds)

Ordinary people narratives in the tv shows everlasting classic

H.Y. The*
Zhejiang Yuexiu University of Foreign Languages, Shaoxing, China

ABSTRACT: Using the narrative approach, this study analyzes ordinary people social roles that contribute to the passing on moral values and good characters. Three narratives of recorded TV show "Everlasting Classic" were used as samples. These narratives described activities of a senior villager, a diaspora, young artists, an opera-singer superstar, and an international artist in performing arts that meant to deliver valuable messages about moral values and good characters written in the classic Chinese books and poems. These ordinary people social engagements in their communities have also inspired the local and international audience to engage in the same activities.

Keywords: narrative, ordinary people, China tv shows, everlasting classic

1 INTRODUCTION

This paper begins here with the Folk Verse of Han, Jiang Nan. Jiang Nan is in the lotus harvest season; Luscious lotus leaves cover the entire pound. Fishes swim cheerfully between the lotus leaves. Those fishes cheerfully move to the East; Those fishes cheerfully move to the West; Those fishes cheerfully move to the South; Those fishes cheerfully move to the North.

Jiang Nan is an area in China that historically presented the prosperous agricultural community around the Yangtze River (Cultural China, 2017). Now, the term Jiang Nan has been used to mean the area around South of Yangtze that includes famous cities such as Shanghai, Anqing, Hangzhou, Nanjing, Ningbo, and Zhenjiang. Carefully selected words and aesthetically used figurative language in the Jiang Nan Folk Verse describe the Han people who lived in China from 202 BC to 220 AD (Lewis, 2007). Using the metaphor of fishes that swim back and forth in the pond, the poet means to tell about the joy of Han people celebrating the harvest festival. The poet also uses lotus, in addition, to show the prosperity of this region. Recurring verses in the folk present the recurring joys of this session that people of Han would always be waited.

With its simplicity, the Folk verse displays people participation in a community activity, the social engagement. When people take part in the community events, their participations can sufficiently be identified as their social engagement (Zhang et al. 2011). The celebrating events certainly also convey the role of citizens and are related to the citizenship issues.

Ordinary people social engagement obviously can be traced using works of art and performance. In order to prove this argument, this study presents the narrative review of three videos from a Chinese TV show, "Everlasting Classic." Celebrities sing the classing Chinese poems in a modern music arrangement, and they share their thoughts about Chinese customs and traditions in this show (CGTN, 2019). The show is a creative process of social re-engineering the classic into modern. Moreover, it is presenting performances and modern visual communication media to juxtapose efforts of citizens to pass down cultural values in non-conventional ways.

This study, thus, means to answer the following questions. First, what are social roles of ordinary citizens that can be identified with the narrative of "Everlasting Classic" shows used as samples of

*Corresponding Author

DOI 10.1201/9781003180128-19

this study? Second, how the "Everlasting Classic" Program has been used by China as an effective media to pass on moral values? Third, what are social engagements and important civic issues that can be identified from the Chinese classic verses presented in the shows?

2 METHODS

The roles of ordinary people in this study were identified using qualitative narrative approach. The narratives were composed by converting the tv shows into written stories. The written stories were then thematically arranged according to the research questions. In the narratives, social roles of the ordinary people were described and explained as cases; it was then connected with the related theories.

3 RESULTS AND DISCUSSION

3.1 *The narratives*

3.1.1 *Grandpa and youth chant the three-character classic*

Zhang Bulang Village, in Henan Province immediately became lively that day. Five stunning boys, members of a Chinese boy band, the Typhoon Teens came to visit Grandpa Zhang Xueli, a senior citizen of the village. The 117-year-old Grandpa Zhang welcomed the boys to his house with cheerfulness and a warm smile. Then, he showed them his birth certificate, recited the entire verses of the Three-Character Classic for the boys, and also showed his calligraphy skill by writing three characters: three, letter, and classic using brush and ink (CCTV, 2019).

The Hunan TV show, "Everlasting Classic" had made it possible for Grandpa Zhang to be aired in a live show without have to do a long-distance travel that may harm his health. During the live show, the TV's host made a video call to Grandpa. Grandpa appeared on the screen, said hello to the program audience and answered the host questions. In the interview, Grandpa also mentioned other Chinese classic books that he had learned when he was a little. Those classics are the Confucian Analects, the Great Doctrine, and the Books of Mencius' teachings. For Grandpa Zhang, the courage to get educated has no boundaries and becoming literate is a gateway for good education. Parts of the classic that are highly related to Grandpa Zhang's good ways of learning are: When young, study; When grown up, perform; Influence the sovereign above; Benefit the people below; Diligence has merit; Play has no advantage; Guard against it; Exert your Strength (Yellow Bridge, 2003).

In the past, teachers and education scholars used various ways to teach literacy and the Three-Character Classic is one of the medias for literacy teaching. This text, that was written by Wang Yinglin, contains not less than 1200 characters, is a mandatory reading for primary school students and it is still being used right now. Because it is easy to understand and universally apply to education around the world, the principles of this teaching further have been adapted by many non-Chinese educations. The text has been translated as well into many languages, including but not limited to English, Spanish, Japanese, and Russia, as well as Indonesian Language.

3.1.2 *Two Chinese opera singers chant ode to the plum blossom*

For the 11-year-old Gabriel Wang, be able to perform on stage with his mentor, a veteran Chinese-Opera-Singer Li Shengsu, was a remarkable achievement and an unforgettable moment. Gabriel and Li deeply impressed the audience with their melodious high-pitched voice singing Ode to the Plum Blossom, a poem written by China's founding father, Mao Zedong (Tay, 1966). Young Gabriel started to have the interest in Chinese Opera Singing when he was 3 years old. He admitted that his interest in singing was aroused through the time he saw and listened to his father practicing (CCTV, 2018).

Gabriel was born in Vancouver, Canada in 2007 and in 2017, he tried to achieve his dream of becoming a professional opera singer by returning to his ancestral-land China and participating in a singing contest. His performance singing the opera attracted the talent recruiters, and soon after that he consistently appeared singing with many well-known singers on TV. As mentioned by his mentor Li Shengsu, for a diaspora of Chinese, at such a young age, be able to sing in Chinese Opera style, Gabriel is a very gifted child. According to Li, she really admires Gabriel for his proficiency in captivating difficult pronouncing words and rhythms as well as his wide range of knowledge about the history of Chinese Opera and Chinese literature.

Gabriel's talent and his charm in Chinese Opera Singing seems to match well with the beauty of spring and spirit of loving the country in the verses of the poem sang as the song with Li Shengsi in "Everlasting Classic." The following is a translation of the poem in English by Tay published in the Journal of Asian Studies in 1966. Ode to the Plum- to the tune Pu-suan-Tzu, Mao Zedong by revising Lu Yu's views in Plum Blossom: "Wind and rain send spring away, Drifting snow brings spring back. The cliff is hanging with thousand-foot ice, A blossom blooms elegant. Elegant but not vying for spring, She only ushers spring in. And when the whole mountainside is in bloom, She (the plum fruit) smiles in the grove [in retirement]."

3.1.3 *A friend from Kazakhstan chants the thousand miles friendship*

Dimash Kudaibergen hailed from Kazakhstan and began singing in China in 2017. He performed initially in the Hunan TV's music program, "I'm The Singer" (Aiyok, 2017). That debut gave him more opportunities to appear in Chinese TV's programs such as Chinese Top Ten Music Awards in Shanghai and Top Chinese Music Award in Shenzhen. Dimash's performance in the Everlasting Classic, aired back after the long break during the pandemic outbreak in 2020, captivated the audience and became hot discussions, both in Chinese and International News. China News Channel, Global Time highlighted Dimash effortless hard work to prepare his singing in the show (Global Time, 2020). The Astana Times (Yergaliyeva, 2020) also quoted that Dimash successfully impressing people who came to the show with his high-pitched vocal range and his sang to a dombra (Kazak's traditional string instrument) accompaniment (CCTV, 2020).

The song that Dimash sang in Everlasting Classic entitled Thousands of Miles, A Common Dream is a lyric written by Kevin based on the Tang era poem, Farewell in Weicheng written by Zhang Jiuling "The ever-budding willows in farewells. The ever-growing clouds in sight. The ever-echoing birdsong in ears. The ever-haunting lovesickness in hearts. The farewell had put us on the gateway. On horse I left for the ferry, seeing you away. Together or apart, near or far. Sharing a heart, friends still we are. In my arms sleeps the moon. Casting velvety dreams to lane and road. The twittering swallow, please. Bestow spring jasmines to every household. Guests from the east and the west. Travelers from the north and the south. Beyond all the mountains and rivers. We are bonded by an angelic soul. Through effort common dreams loom. In every home sprint will bloom." [Fandom Wiki, n.d]. Both Dimash singing of this song and Kevin's work on recomposing the verses have added nuances and emotions to the original poem. Their collaboration also shows that people from different countries are different and at the same time can develop friendship and have a common dream.

3.2 *Analysis of the narratives*

3.2.1 *Social roles and civic engagement*

The term "ordinary people" in this study is used to describe people that are not public officials, not take on leadership roles, and not serve in the military. Simply, they are civilians with various social roles. These ordinary people carry out their daily routines according to their respective professions. In these daily routines, they are capable, enthusiastic and qualified to govern their lives to inspire others (Welzel & Inglehart, 2008). Daily routines of these empowered civilians can be referred to as social responsibilities and civic engagement that have been extensively studied by researchers in the field of Civic Education.

Grandpa Zhang, as a senior citizen, shows the young people that age is not a restriction to take on social involvement. Typhoon Teens are the representation of grown ups who still learn to be responsible citizens. The event of these youths visiting Grandpa tells about respecting and loving the elderly. Today, numbers of senior citizens in China have been increasing rapidly and some of them have encountered challenges related to personal and social empowerment (Lai & Ruan, 2020). The elderly and youth harmonious relationship is an important civic issue in China today. The harmonious relation has been proved supporting the senior citizens physical and psychological well-being (Tang et al. 2018).

Another important issue of citizenship in China is the overseas diasporas. The social role of Gabriel Wang as a Chinese diaspora here can become an ideal role model of hard work and enthusiasm for other Chinese children to develop their potential talents. His performance in the Everlasting Classic implicitly has a message about the importance of adoring and learning from the idol and inspiring others by becoming their idol. Gabriel's mentor, Li showed a good example of democratic education. Instead of just teaching and telling, she emphasized that youths like Gabriel need to be accompanied to pursue their careers. Many Chinese diasporas with professional talents like Gabriel have returned to China today to pursue their careers. The Chinese diasporas have benefited China in two ways: strengthening the country's domestic development and increasing the country's international geostrategic (Wang, n.d.). For the diaspora themselves, the opportunity to develop a career in China not only economically benefits them but also culturally pays off because they can embrace the connection between their identity to their ancestral Chinese background (Hutton, 2018).

The social role of Dimash is also very important to our globalized world today. He is a representation of world citizens who are willing to crossing the border of their countries to expand carriers. In addition, he is also carrying the message about continuous learning from other communities.

Activities of these people are very common and universally appear in our daily lives; However, not until TV programs or news featured them, the civic roles of ordinary people mostly go unnoticed. These people supposedly are not being in the position of unnoticeable in our societies. Less or more, ordinary citizens play important roles in carrying out the issues of civic and citizenship. When the living stories of ordinary people are portrayed as an entertainment program such as the "Everlasting Classic," messages about the social roles of citizens become easier to take on for other ordinary people. The message will somewhat influence the change of behavior and lifestyle of others. For that reason, Civic Education Scholars have to learn from journalists about the inclusion of writings as well as academic discussions about ordinary people in order to empower their positive and actions to influence many other ordinary people in order to take progressive actions that can benefiting the entire communities (Welzel & Inglehart, 2008).

It can be argued that scientific seminars, academic discussions, class lectures or lessons with the aim to instill good characters and moral values are less influential than the entertaining ways in achieving the same goal. The language that academics use to send the message is not as simple as the language of entertainment. Technical terms that academics used in textbooks, papers, and presentations frequently were not in the context that easy to understand by ordinary people (Plaven-Singray et al. 2017); Differently, the entertainment is using attractive ways, such as semiotics (signs and symbols) that are easy to see and simple terms, body language, and other visual designs to engage people to the show (Gaines, 2010).This is obviously a challenge to academics, especially in the field of Civic Education for finding a way to be as attractive as the entertainment or perhaps take this entertaining way into account when designing the educational programs.

3.2.2 *Creative entertainment as the media to pass on culture*

As an entertainment program, The Everlasting Classic creatively unveils the history, knowledge and moral values in the Chinese classical scriptures. It involves celebrities adapting verses in the scriptures, making adjustment for the purpose of creative expression, and singing the songs. Rearrangement for the purpose mentioned above enabling actualization of moral values and context of living in the past to current time without losing the important messages of the verses' writer. In addition, singers have opportunities to express the learning process of their creative arts preparation stages till the end of their performance. Occasionally, juries of the program and the host request the artists to perform their additional skills and their knowledge about history as well as values related to their performance.

The international artists and Chinese diasporas also take part in the Everlasting Classic. Their performance sends an important message that performance arts have universal values. Apart from being an entertainment, Everlasting Classic can help people to learn Chinese history, culture and values, as well as literature and performing arts. This program is also a place for artists to learn from one another, develop their friendship and strengthen the collaboration with performing together at one stage.

The Everlasting Classic performance-stage looks appealing; It has modern digital-visual design. Judges and audiences in the auditorium could vote for their favorite artists immediately from a hand-touch sensor that has been modeled as a heart and attached to their clothes before entering the performance studio. Viewers on the TV could follow and vote for their idols using the QR-code showed in the screen. Artificial lighting in the shape likes bamboo-blades is also used to show verses and other characters of people who are watching the show. The stage also has various movable decorations that can be rearranged to suit different performances. In addition, the artists also occasionally wear stunning ethnic outfits and brought traditional instruments to accompany their performance. The visual and art designs have rich semiotics (signs and symbols) that function effectively to accompany body languages of the artists and textual language of the verses to deliver entertaining and meaningful messages (Atoofi, 2015).

3.2.3 *Samples of social engagement and moral values embedded in Chinese classic*

The term "Classic" attached to Chinese books and poems indicates the long history of the scriptures. These scriptures stored important wisdom of the past that are still up-to-date to our societies today. Therefore, not very difficult to find our lives today connected to the living of people in the pass that were documented in the texts. Chinese people in the era of Song, for example have been engaging in practicing the educational principles written in the Three-Character Classic. The Three-Character Classic consists of five important topics: 1) goodness of mankind and social relationships; 2) basic principles of Confucius's thought; 3) history of China; 4) exemplary behavior; and 5) encouragement for learning. This text has great influence to the development of public education that can provide people with opportunity to learn regardless their social status in the society (Liu, 1985). The text is not only a literary resource, but it is first the source of Civic Education for Song's people and later for the Chinese. Then, when it comes to the society where the text has significant influence, it has become the guide for good moral characters and roadmap for becoming responsible citizens.

The universal values of the moral teachings related to the narrative about Grandpa Zhang and the Typhoon Teens can be identified in the section about study and hard work. The verse stated that learning is the foundation for growth and must be accompanied by continuous practice till mastering the skills; A person who is successful in applying his learning outcomes to become a useful human being will benefit others for generations to come. Another verse emphasized the need to set limits of having fun in life; Do things as needed and avoid overdue-ness; Self-control and discipline are necessary for life. What do you think about these two verses? Do you agree that the moral teaching like this is very universal and can be found in most societies? Do you also agree that moral values are still very actual in guiding us in carrying out social responsibility and involvement in society?

Poems as a medium for communication have been used for many purposes, including persuasion, encouragement, and motivation (Trench, 1930). Mao Zedong certainly is not the only leader to use the poem to boost the spirit of civilians' toughness and patriotism. Indonesia's founding father, Soekarno also had used the poem entitled *"Aku Melihat Indonesia"* (Ahmad, 2020) with similar aim to Mao, making civilians proud of Indonesia. Both Mao and Soekarno have proved that poems are not only words and verses; poems are a very effective tool and strategy to influence "ordinary people" following their lead fighting the imperialists and to stand out as a free nation state (Hagemann, 2006). Literary scholars noticed that Mao's poem was indeed the remake of Luo You's poem with the same title but for a very different purpose. At the time Mao wrote this poem, 1961, China was in a very serious economic recession. Mao aims to awaken civil resilience and nationalism using the poem by illustrating the living conditions of China as the plum tree. Mao wants people to see the crisis as the extreme weather that will be over. The resilience is similar to the toughest plump tree to face extreme weather. It will bloom and the nation will be back to glory if civilians hold strong together to face the crisis.

The last classic text to discuss here is a piece of work written during the Tang Dynasty, the golden era of literature in the history of ancient China. This poem is about friendship and the poet writes this for a friend who was leaving for a journey. Although the poem was primarily aimed for personal communication but this personal aim strongly connected to the experience of many people. Many people will be experiencing the time for setting apart and living people as well as the place they love. The journey will then remind people of distances and boundaries that can separate us physically but not mentally. Distances and boundaries cannot disconnect the feeling of acquaintance and the memory of times people spent together. Distances and boundaries are just a way to make human beings realize about hopes and dreams. Obviously, verses of this poem are about the living experiences of us wherever we are as human beings.

4 CONCLUSIONS

Using the narrative approach, this study revealed the social role of ordinary citizens, the function of the performance in conveying noble values and good characters, and the citizenship issues embedded in the classical books and poems. The social roles that can be identified are senior villagers, teenage artists, Chinese diasporas, Chinese opera song superstars, and international artists who are developing artistic careers in China. These people appear as main roles in the video recordings. Their stories captivate the audience interest, engage them to watch the show, and inspire them to do the similar social engagements. The visual design semiotics of stage and the broadcasting footage are two important factors to affect the effective messages meant to deliver to the audience in addition to the artist's performance. These physical features facilitated the persuasion simultaneously with the support of the media and the artists performances. Most importantly, classical Chinese books and poetry have universal values that can be connected with the lives of other peoples of other nations. These universal values include study hard, work hard, love and be loyal to the homeland, as well as appreciate friendship.

REFERENCES

Ahmad, T. 2020. *Aku melihat Indonesia, puisi karya Presiden Soekarno Ini ungkap gelora Pantai Ngliyep.* JatimTimes

Aiyok, 2017. Dimash'es all performances in "The Singer 2017". Retrieved from: https://www.youtube.com/watch?v=t2Dhil2n59o.

Atoofi, S. 2015. Context from a social semiotic perspective: a discourse analytical study of the children TV show Bubble Guppies. *Social Semiotics* 25(5): 558–577.

CCTV China Central Television. 2018. *(Classic chanting) Li Shengsu and the 11-year-old boy wang hongxiang sang Ying Mei in Beijing Opera.* Retrieved from: https://www.youtube.com/watch?v=iiW85n3aMO8.

CCTV China Central Television. 2019. *(The second season of classic chants) The 117-year-old man and the Typhoon Youth Group sang the Three Character Classic in a cross-century chorus.* Retrieved from: https://www.youtube.com/watch?v=zIkoyLARdG8. (Accessed on February 20, 2020).

CCTV China Central Television. 2019. *The second season of classic ching. xiamen sixth middle school choir sings fish and lotus leaves, the most beautiful interpretation of Jiangnan.* Retrieved from: https://www.youtube.com/watch?v=feE9Y_bbahs.

CCTV China Central Television. *2020 Dimash kudaibergen, A dream of one heart.* Retrieved from: https://www.youtube.com/watch?v=0LziF8Mty58.

CGTN. 2019. *Variety show 'everlasting classics' gives antique chinese literature a pop tune.*

Cultural China. 2017. Jiang Nan: South of the Lower Reaches of the YangTze River.

Fandom Wiki. n.d. Thousands of miles, a common dream. Retrieved from: https://dimash.fandom.com/wiki/Thousands_of_miles,_a_common_dream. (Accessed on June 4, 2020).

Gaines, E. 2010. Media Literacy and Semiotics. New York: Palgrave Macmillan.

Global Time. 2020. 'Everlasting Classics' back on air after hiatus. Retrieved from: https://www.globaltimes.cn/content/1192838.shtml. (Accessed on May 1, 2020).
Hagemann, K. 2006. "Be proud and firm, citizens of Austria!" Patriotism and masculinity in texts of the 'political romantics' written during Austria's Anti-Napoleonic Wars. German Studies Review 29(1): 41–62.
Huiyao, W. n.d. *China's return migration and its impact on home development.* UN Chronicle
Hutton, J. 2018. *Journey to the past: Why overseas Chinese are finally embracing their roots.* South China Morning Post.
Lai, D.W.L. & Ruan, Y. 2020. Revisiting social work with older people in Chinese contexts from a community development lens: When east meets west. *Community Practice and Social Development in Social Work:* 245–261.
Lewis, M.E. 2007. *The early Chinese empires: Qin and Han.* Cambridge: Harvard University Press.
Liu, J.T.C. 1985. The classical Chinese primer: Its three-character style and authorship. *Journal of the American Oriental Society* 105(2):191–196.
Overseas Chinese Language and Education Online Chinese Culture Tutorial Network. 2019. *Hanyuefu. Yuefu Songs of the Han Dynasty.*
Plavén-Sigray, P et al. 2017. The readability of scientific texts is decreasing over time. Elife, 6, e27725.
Tang, F. et al. 2018. Employment and life satisfaction among middle- and old-aged adults in China. *Gerontol Geriatr Med* 4: 1–8.
Tay, C. N. 2020. From snow to plum blossoms: A commentary on some poems by mao Tse-Tung. *The Journal of Asian Studies* 25(2): 287–303.
Trench, W. F. 1930. The function of poetry according to Aristotle. *Studies, Irish Quarterly Review* 19(76): 549–563.
Welzel, C. & Inglehart, R. 2008. *Democracy as human empowerment: The role of ordinary people in the emergence and survival of democracy.* UC Irvine: Center for the Study of Democracy
Yellow Bridge. 2003. *The three-character classic: Confucian roadmap for kids- paragraph 5: closing statement.* Retrieved from: https://www.yellowbridge.com/onlinelit/sanzijing5.php
Yergaliyeva, A. 2020. *Pop Kazakh singer Dimash kudaibergen performs on Chinese central television to raise support Wuhan medical workers.* Retrieved from: https://astanatimes.com/2020/06/pop-kazakh-singer-dimash-kudaibergen-performs-on-chinese-central-television-to-raise-support-wuhan-medical-workers-video/
Zhang, S. et al. 2011. Integrating online and offline community through Facebook. *International Conference on Collaboration Technologies and Systems (CTS):* 569–578.

Empowering Civil Society in the Industrial Revolution 4.0 – Sukowiyono et al (Eds)

Multicultural education through field-based learning at water sites

M.N.L. Khakim*, W.D. Sulistyo, M.N. Jauhari & N.H. Marsudi
Universitas Negeri Malang, Malang, Indonesia

ABSTRACT: Multicultural education is a solution to prevent the threat to a national division. In multicultural education students are taught to aware and responses positively to the diversities of the nation. In order to have the multicultural education implemented, this research examines the field study of local history at the spring site in the Singosari-Lawang. From visiting the site, students had opportunities to observe closer the lessons of history they learned in class. This study showed that students were able to understand that this plurality has existed on the Majapahit era, where even though the people adhere to different religions, they can still coexist peacefully and not drop each other's beliefs.

Keywords: multicultural education, field-based learning, springs

1 INTRODUCTION

Indonesia is inhabited by people with various ethnicity, different cultures, thousands of languages, and many religions; creating the nation as a pluralistic society (Kamal, 2013). Indonesia as a very diverse country has the potential to be divided based on the characteristics of the community diversities. The inequality in granting rights to minorities can cause the intention of separation (Rudianto, 2012; Suparlan, 2004).

One of the ways that can be done in order to create a peaceful-united plural society is through multicultural education. Multicultural education can be understood as the education about cultural diversity of the community and how to respond to the diversity (Banks, 1993; Bennett, 1986). The objectives of multicultural education include helping students to see the world from various perspectives, equipping students with knowledge about other cultures (other ethnic cultures or other religions), reducing discriminatory attitudes or behaviors against other races and religions. and the last is helping students to be able to master how to read, write and count (Banks, 2008; Modood, 2007).

By studying the history of spring sites in the Singosari-Lawang area, students can acquire knowledge about the diversity of Indonesia in the time of Majapahit. The knowledge is embedded in the story of King Hayam Wuruk, the Hindu's King building Stupa Sumberawan in a site of people worship Buddha (Muljana, 1979; Poerbatjaraka, 1924). In addition, students can also take lessons from the culture and conditions on other sites. From these stories and cultures, it is hoped that students will have a sense of tolerance religious diversity of Indonesia.

Students learn by visiting the historical sites. Students can witness historical relics directly from this field study (Gilbertson et al. 2006). Students, in addition, learn to improve social, cooperation and good communication skills (Mygind, 2007). The experiences that students got from visiting the site can help them develop respect to history and cultural inheritance.

---

*Corresponding Author

 DOI 10.1201/9781003180128-20

2 METHODS

This study applied the descriptive-qualitative approach to collect and analyze data (Creswell & Creswell, 2017; Semiawan, 2010). Primary objects of this study are four sites located in the Singosari area to lawing: Patirtan Watugede, Polaman, Mbiru and Sumberawan. Data was collected by doing observation, documentation, and interviews. The researchers observed an interviewed security guard of the site and the community around the site. In order to analyze the data, we also carried out by literature study and reviewed previous research.

3 RESULTS AND DISCUSSION

3.1 *Spring sites in the Singosari-Lawang area*

From the spring sites of Singosari-Lawang area, this study found historical information, their condition and multicultural values contained in the site. The findings are presented in the table 1 below.

Table 1. Site presentation as a learning resource

History	Utilization	Multicultural Values
Sumberawan – Singosari		
Built by King Hayam Wuruk who is a Hindu, ordered his subordinate to build the Sumberawan Stupa, whose no-taben is a place of worship for Buddhists. (Muljana, 1979; Poerbatjaraka, 1924)	Used by Buddhists and Hindhu for worship. The difference is, if Buddhists use the Stupa as a means of worship, Hindus take water from a source and use it for worship on Vesak days. Other uses for irrigation by the community	Multicultural values can be seen from the story of Raja Hayam Wuruk, a Hindu who built a place of engagement on Buddhists. Apart from that, Hindus and Buddhists alike still use Sumberawan as a place of worship. This proves the existence of harmony between religious communities in Indonesia.
Polaman – Lawang		
The name Polaman is believed to be derived from the words *pa* and *ulaman* which mean a place to keep fish (Damayanti et al. 2017; Wurianto, 2012). In this place there are various types of fish. It is said that the Polaman water source was a stopover for Kadiri kings who traveled to the eastern area of Mount Arjuna and Mount Kawi. (Hardjowardojo, 1965).	On this site, there are adjoining places of worship for Hindus and Muslims, namely a mosque located to the south of the pond and a small temple located to the west of the pond.	Harmony between religious communities is also seen in Polaman. It can be seen from the existence of a mosque and a small purée in the location of the source of Polaman, which shows that there is more than one religion, namely Islam and Hinduism, which each use of water source for worship.
Mbiru – Singosari		
Negarakertagama tells the story that King Hayam Wuruk on his pilgrimage took the time to enjoy the view of the *kedung mbiru* (Muljana, 1979; Poerbatjaraka, 1924). The function of this spring used to be a place to purify pilgrims before going to the top of the sacred mountain Sang Hyang Arjuna which is a representation of the peak of the holy mountain Sang Hyang Mahameru. (Firmansyah & Soesilo, 2018).	Currently, the function of this water source is as a place for Javanese people to perform rituals on certain days. According to the ritual performers, they took water, for bathing, repelling balak, to launch a fortune.	The ritual performed on certain days by the Javanese Muslim community is an acculturation of local and immigrant beliefs. From there we can take action in the form of not intimidating each other's beliefs, because they may still be closely related to the beliefs we have.

(*Continued*)

Table 1. Site presentation as a learning resource (*Continued*)

History	Utilization	Multicultural Values
Watugede – Singosari		
In history, this petirtaan was a place for the purification of the king's daughters, especially a very beautiful princess named Kendedes so that people knew the place as Patirtaan Kendedes (Junianto & Subadyo, 2017). Meanwhile, Benyak also knows as Wa-tugede according to the location of the tirtaan. But actually the name patirtaan is boboji which means to be holy.	Apart from being used as a tourist spot for bathing/swimming, now patirtan Ken Dedes is also used as a place for certain rituals by Hindus, especially Javanese Hindus. These rituals include the ritual of taking 'holy water' from a water source, the ritual of 'soaking' in a patirtan pool and the ritual of bathing in a water source. (Junianto & Subadyo, 2017).	Sometimes due to different beliefs, we perceive other people's beliefs as wrong. Like the rituals carried out by Javanese Hindus in Patirtan Kendedes. Sometimes other people misunderstand the goal. Whereas the Javanese Hindus do not worship water or other objects in the patirtan kendedes, but use them as a medium to worship God.

3.2 *Site based learning activities of springs*

A field study is one of the many learning models. Basically, a field study is an activity designed and developed by teachers to combine learning in the room and outdoors by utilizing the surrounding environment or open nature (Hana & Sulistyo, 2020; Knight, 2013; Sulistyo, 2019). In addition, a field study allows students to actively explore the surrounding environment in order to achieve three aspects of learning skills, namely cognitive, affective and psychomotor skills (Nisa, 2015; Woodhouse & Knapp, 2000).

The field study is not just studying or discussing outside the room, but rather as an application of everyday life (Burriss & Burriss, 2011; Hasan, 2016). This is closely related to the mandate contained in the 2013 Curriculum; students must have skills. The skills referred to in the 2013 Curriculum cover many things, from knowledge, attitudes and practices in the field. In the field learning based on spring sites in this region, it is carried out in three main stages, namely first, the initial activity with an introduction or orientation process about the history of the site. This activity begins with the formation of groups and small-scale group research on the Sumberwan spring site which is conducted in the classroom. The second activity which is the core activity is carried out by the process of observation and exploration carried out by students at the Sumberawan spring. In this activity students are required to find as much information about spring sites in the Singosari-Lawang area (including Sumberawan, Polaman, Mbiru and Patirtan Kendedes) as much as possible. Observation and exploration activities were also carried out to clarify findings and results from small-scale research conducted in previous classes. The third activity is to present and discuss the findings from the research and exploration of each group. The third activity focuses on information exchange and value reflection activities. One of the success of the learning process is determined by the learning planning factor that is in accordance with the implementation (Sulistyo et al. 2020; Sulistyo & Wiradimadja, 2019).

3.3 *Multicultural education*

Multicultural education can be defined in many ways. Multiculturalism education is education that pays special attention to developing an attitude to tolerance and respect for cultural, ethnic and religious differences and also granting equal rights to minorities (Kymlicka, 2007). Multiculturalism education can be also being defined as education in order to create respect for the diversity of ethnic and community cultures (Tilaar & Hapsari, 2004). In addition, Multicultural education is education that teaches students to respect the cultural diversity of society (Rosyada, 2014). From some of the above definitions it can be concluded that the notion of multicultural education is education that teaches an attitude to tolerance towards a diversity that exists in a society. In this learning, students are emphasized on diversity in matters of religion. As we know, Indonesian society adheres to five different religions which make it very vulnerable to divisions. Therefore, Multicultural Education is deemed very important to prevent this.

The purpose of this learning is to instill an attitude to religious tolerance of students. This tolerance is one of the many character values that must be instilled in students and from a good learning process we can instill character (Sulistyo, 2016; Sulistyo & Pamungkas, 2020). This attitude to religious tolerance can be taken from the behavior of King Hayam Wuruk who was a Hindu, giving orders to build the Sumberawan Stupa which is a Buddhist practice (Muljana, 1979; Poerbatjaraka, 1924).

This also shows that since the Majapahit era, multiculturalism has been created in the archipelago. There have been two religions that are side by side and also peaceful. Do not collide with each other's arguments to show which religion is the best and most true, but to help each other and respect each other's beliefs. In addition, students can also take the value of multiculturalism from the conditions and functions of the existing spring sites in the Singosari-Lawang area, including the mosque and small puree adjacent to the Polaman site, depicting harmony between religious communities and also the provision of rituals in the area. Mbiru and Watugedhe sites that teach students about respecting cultures from other religions. With this knowledge of multiculturalism, it is hoped that students will have a high sense of religious tolerance so that they do not have racist attitudes to other beliefs. In addition, students are also expected not to be easily consumed by negative issues that have the aim at dividing the nation.

4 CONCLUSIONS

The biggest challenge of a pluralistic nation is the threat of national division. Therefore, it takes several efforts to prevent this. Outdoor learning based on using spring sites as a learning resource is an effort to prevent the threat of national division. Based on some historical information in the spring site in the Singosari-Lawang area that students have explored, it is hoped that students can take lessons of pluralism and tolerance in the archipelago. The field study activities are carried out so that students can directly observe the site conditions. This activity is also carried out to clarify findings obtained from books or other written sources. Outdoor learning activities are divided into three main stages, namely orientation activities, exploration activities and presentation and reflection activities. The purpose of this study is to instill tolerance in students from the history of the spring site in the Singosari-Lawang area. From this learning students were able to understand that this plurality has existed on the Majapahit era, where even though the people adhere to different religions, they can still coexist peacefully and not drop each other's beliefs. This of course refers to multicultural education which aims to respect and appreciate differences and equalize the rights of the majority of minorities.

REFERENCES

Banks, J. A. 1993. Multicultural education: Historical development, dimensions, and practice. *Review of Research in Education* 19(1): 3–49.
Banks, J. A. 2008. *An introduction to multicultural education*. New York: Pearson.
Bennett, C.I. 1986. *Comprehensive multicultural education: Theory and practice*. Boston: Allyn and Bacon.
Burriss, K. & Burriss, L. 2011. Outdoor play and learning: Policy and practice. *International Journal of Education Policy and Leadership* 6(8), 1–12.
Creswell, J.W. & Creswell, J.D. 2017. *Research design: Qualitative, quantitative, and mixed methods approach*. Sage Publications.
Damayanti, F. et al. 2017. Ruang budaya "Barikan" di Desa Sumber Polaman Lawang Jawa Timur. *RUAS (Review of Urbanism and Architectural Studies)* 15(1): 48–66.
Firmansyah, D. & Soesilo, F. 2018. *Sejarah singkat Kecamatan Singosari dan mengenal tinggalan kesejarahannya: dari masa prasejarah sampai masa kemerdekaan*. Malang: Inteligensia Media.
Gilbertson, K. et al. 2006. *Outdoor education: Methods and strategies*. Human Kinetics.

Hana, S.I. & Sulistyo, W. D. 2020. *The utilization of colonial government buildings in malang as the learning sources*. 404: 278–282.
Hardjowardojo, P. 1965. *Pararaton*. Bhratara.
Hasan, Z. 2016. Upaya meningkatkan kualitas pembelajaran sejarah Indonesia kelas XI IPS 1 melalui penerapan outdoor learning berbasis inkuiri di SMAN Kunir tahun ajaran 2015-2016. *Sejarah dan Budaya: Jurnal Sejarah, Budaya, dan Pengajarannya* 10(2): 209–215.
Junianto, R.N.S.L. & Subadyo, A.T. 2017. Revitalisasi situs patirtan Watugede Singosari sebagai obyek wisata spiritual berkelanjutan. *Seminar Ikatan Peneliti Lingkungan Binaan Indonesia (IPLBI) 1*.
Kamal, M. 2013. Pendidikan multikultural bagi masyarakat Indonesia yang majemuk. *Al-Ta Lim Journal* 20(3): 451–458.
Knight, S. 2013. *Forest school and outdoor learning in the early years*. Sage.
Kymlicka, W. 2007. Multicultural odysseys. *Ethnopolitics* 6(4): 585–597.
Modood, T. 2007. Multiculturalism. *The Blackwell Encyclopedia of Sociology*, 1–4.
Muljana, S. 1979. *Nagarakretagama dan tafsir sejarahnya*. Bhratara Karya Aksara.
Mygind, E. 2007. A comparison between children's physical activity levels at school and learning in an outdoor environment. *Journal of Adventure Education & Outdoor Learning* 7(2): 161–176.
Nisa, J. 2015. Outdoor learning sebagai metode pembelajaran ips dalam menumbuhkan karakter peduli lingkungan. *Sosio didaktika: Social Science Education Journal* 2(1): 1–11.
Poerbatjaraka, R.N. 1924. Negarakertagama. *Bijdr. TLV, Deel LXXX*.
Rosyada, D. 2014. Pendidikan Multikultural Di Indonesia Sebuah Pandangan Konsepsional. *Sosio Didaktika: Social Science Education Journal* 1(1): 1–12.
Rudianto, Y. 2012. Fenomena kekerasan sosial dan struktur majemuk masyarakat Indonesia. *Jurnal Administrasi dan Kebijakan Publik*, 1(1): 67–92.
Semiawan, C. R. 2010. *Metode penelitian kualitatif*. Jakarta: Grasindo.
Sulistyo, W.D. et al. 2020. Learning experience from learning sources: Exploiting geographic and historical potential of guerrilla sites in Wonokarto Pacitan as a source of historical learning. In *IOP Conference Series: Earth and Environmental Science*.
Sulistyo, W.D. 2016. *Pengembangan model pembelajaran sejarah berbasis nilai-nilai serat wicara keras untuk meningkatkan kemampuan berfikir kritis siswa sma insan cendekia al mujtaba sukoharjo.* (Unpublished Thesis). Surakarta: Universitas Sebelas Maret.
Sulistyo, W.D. 2019. Learning activities from learning resources: Pemanfaatan dan pemaknaan situs sejarah kawasan alun-alun merdeka kota Malang. *Jurnal Pendidikan Sejarah Indonesia* 2(1): 49–63.
Sulistyo, W.D. & Pamungkas, O.Y. 2020. Pemanfaatan situs sejarah peradaban Islam di kota Malang sebagai aktivitas belajar untuk menanamkan nilai karakter. *Agastya: Jurnal Sejarah Dan Pembelajarannya* 10(1):1–9.
Sulistyo, W.D, & Wiradimadja, A. 2019. Lesson Study (LS): Memahamkan "masalah penelitian" kepada mahasiswa. *Jurnal Teori dan Praksis Pembelajaran IPS* 4(1): 29–37.
Suparlan, P. 2004. Masyarakat majemuk, masyarakat multikultural, dan minoritas: Memperjuangakan hak-hak minoritas. In*Workshop Yayasan Interseksi, Hak-Hak Minoritas dalam Landscape Multikultural, Mungkinkah di Indonesia?*
Tilaar, H.A.R. & Hapsari, S.D. 2004. *Multikulturalisme: Tantangan-tantangan global masa depan dalam transformasi pendidikan nasional*. Jakarta: Gramedia Widiasarana Indonesia (Grasindo).
Woodhouse, J.L. & Knapp, C. 2000. *Place-based curriculum and instruction: Outdoor and environmental education approaches*. Charleston: ERIC Clearinghouse on Rural Education and Small Schools.
Wurianto, A.B. 2012. Aspek budaya pada upaya konservasi air dalam situs kepurbakalaan dan mitologi masyarakat Malang. *Jurnal Humanity* 4(2): 21–32.

Empowering Civil Society in the Industrial Revolution 4.0 – Sukowiyono et al (Eds)

Actualization of Mataraman society virtues on life

N.W. Rochmadi*
Universitas Negeri Malang, Malang, Indonesia

ABSTRACT: Mataraman is the term for the Javanese elite community, which models for the Javanese character. This research aimed to describe the actualization of the virtues in life. The research was conducted with a descriptive qualitative research design. Data collection techniques used observation, interviews, and study documentation. Data was analyzed using the interactive model. Research findings were (1) the mataraman society virtues were: (a) simplicity, (b) respect for others and parents, (c) introspection, *mawas diri, nandhing sariro, ngukur sariro, tepa saliro, mulat saliro*, (d) *gemi, nastiti, ati-ati*, (e) harmony, (f) surrender to God, (g) wisdom, and (h) humbleness; (2) actualization of the virtues in life with language, use of words, volume and intonation, behavior (*gerak laku*), clothing (*busana*), leadership in the family, and traditions. Based on the research findings, be suggestions: (a) make the virtues as a guideline in behavior; (b) used to behaving in accordance with the virtues.

Keywords: actualization, virtues, Mataraman

1 INTRODUCTION

Human life cannot be separated from groups, society, nation, and state. Through groups, humans express and try to realize their ideals in life, one of its forms is by creating a culture. Culture is a reflection of human life in reality and ideals.

Human cultural life in social, national, and state activities is manifested based on ideal values or virtues. Therefore, every community always has virtues, which serve as guidelines for their behavior in realizing a better life. The virtues of the community are manifested in the form of various rules of life in society, in the form of norms that become the order and reference in interacting with others and with the environment.

Each individual is controlled by the norms (*rules*) of the society, nation, and state. Basically, the existence of norms in human life aims to protect and care for faith or belief in God, respect and appreciate individual values, which are honor and must be respected by everyone, sustain life, carry on offspring and maintain ownership which is a gift of God earned through work.

Humanity in culture is manifested in the form of social, national, and state activities, through a social contract or collective agreement, referring to the virtues of the nation and norms. Being a society and nation and state are human instincts, as social beings, and for people who are religious, at once are orders from God. However, different from the gathering of animals, humans come together and create a better life by referring to virtues such as justice, security, welfare, deliberation, and peace. In the context of Indonesia, these virtues are formulated in a value system consisting of five principles, called Pancasila, namely *Ketuhanan Yang Maha Esa, Kemanusiaan yang Adil dan Beradab, Persatuan Indonesia, Kerakyatan yang dipimpin oleh Hikmat Kebijaksanaan dalam Permusyawaratan/Perwakilan dan Keadilan Sosial bagi seluruh Rakyat Indonesia.* On the basis of these five principles, all Indonesian people unite and live together on the basis of the principle of kinship within the unitary state of the Republic of Indonesia.

*Corresponding Author

DOI 10.1201/9781003180128-21

Be society and state are regulated by an authority established through noble agreements and agreements in the form of norms. Regulations in the form of norms are aimed at ensuring the achievement of good and absence of bad and evil in life together, as well as being the mission of the state authority.

As in state life, social and national life also refers to virtues that are shared in their truth. These virtues become a reference in shaping the order of life in society and as a nation and state. The virtues and order become a reference in the behavior of the supporting community in interacting with others and with their natural environment. The ownership of behavior in accordance with the values and order of society is what ultimately becomes the identity of the communities concerned, one of which is the Mataraman community.

Based on the Javanese structure, Javanese ethical values serve as a guide for Javanese people when they do activities and interact with fellow humans in the life of the state and society. These Javanese ethical values are contained in an order known as the *unggah-ungguh*. The Javanese ethical values and *unggah-ungguh* are used as guidelines for behavior when interacting with fellow humans, and the environment.

Mataraman is a portrait of the ideal and role model of Javanese society, although among them there are still further groupings and have various characteristics. However, they are a group of Javanese people who are consistent in obeying and implementing Javanese ethical values and structures. Based on this, an exploratory study was conducted on the virtues of the Mataraman community and their actualization in they life.

2 METHODS

In accordance with the objectives, this research was conducted with a descriptive qualitative design. The researcher conducted an exploratory study on the actualization of the virtues of the Mataraman community in Blitar City. The data sources of this research are those who understand the virtues of the Mataraman community in the Blitar City, including Sujilah, Totok, Mardiono, Purnaningsih, Willis Purwanto, Setyo Winarno, Sutadi and Dyah Palupi. Data collection was carried out by observation, interviews, documentation, and focus group discussions. The data analyzed refers to an interactive model from Miles and Hubermans.

3 RESULTS AND DISCUSSION

3.1 *The virtues of the Mataraman society*

The Mataraman society is a Javanese community that socially and culturally has a relationship with the influence of the Kerajaan Mataram culture, from the Hindu-Buddhist era to the Mataram Sultanate era which was centered in Yogyakarta and Surakarta. The existence of the mataraman community around the city of Blitar is a necessity. The existence of the Mataraman community in the city of Blitar is related to the history of the Javanese war (De Java Oorlog), the war against the Dutch (VOC) led by Pangeran Ontowiryo (Pangeran Diponegoro) between 1825–1830 (Moedjanto, 1987).

The defeat of Pangeran Diponegoro in the Java war resulted in those who became Laskar Diponegoro fleeing from the Yogyakarta area to all corners of Java, including to Blitar, by disguising themselves, opening new hidden land, staying as *magersari* (hitching a ride in someone else's yard or living land owned by the state while cultivating the land). Laskar Diponegoro, who fled to Blitar, grew and developed to form its own villages in several areas around the city of Blitar, the area called Kampung Mataraman, said Mr. Mardiono.

Based on the Javanese norms, Javanese ethical values should serve as a guideline for Javanese people when they do activities and interact with fellow humans in life, as well as in society. These Javanese ethical values are contained in an order known as *unggah-ungguh*. Javanese humans who implement Javanese virtues are given as *wong mbeneh* or people who have virtues, if their behavior is not in accordance with these values then it is called *wong ora mbeneh*. The statement is in accordance with that delivered by Mr. Mardiono, "*Tiang jawi meniko kedhah mbeneh. Tiang jawi meniko kedah manut kaliyan ugi ngikut dateng tatanan lan budaya jawi*".

Likewise, what was conveyed by Mr. Totok as a descendant of Mataraman and a teacher by profession, "Javanese people, especially Mataraman, must be good, have virtues. His daily behavior in interacting with fellow humans must be in accordance with the Javanese culture and order." He further said that "If humans can behave in accordance with Javanese culture, Javanese virtues, it is believed that they will find *lahir batin* safety in their lives, both themselves and their families".

According to Mr. Mardiono, "The characteristics of the Javanese Mataraman are simple, harmonious and introspective, including *nandhing sariro, ngukur sariro, tepa saliro, mulat saliro.*" Meanwhile, Mr. Totok stated that "Mataraman people have high respect for parents, are simple, wise, humble and not arrogant, it can be seen from the house, simple from the outside but luxurious inside. In addition, Mataraman people usually have a high surrender to God."

Meanwhile, according to Mrs. Sujilah, "The mataraman people usually even *nopo wontene, atos-atos uripe, rukun kalih liyane, khurmat kaliyan sapodho-podho* have a sense of *wedi, isin, sungkan* to others; fear of prayer; *takut kuwalat; andhap asor; ngapuracang; sendiko dhawuh; sabar lan narimo, pasrah lan sumeleh; percaya akan pepesthen, jodho, rejeki, pati garise pengeran. eling sangkan parane dumadi*".

The above statement was approved by Mrs. Purnaningsih (wife of the Mataraman resident), "*Bapak kulo meniko dalang, saben-saben tansah maringi pitutur luhur; dedalane guna lawan sekti iku kudu andhap asor, wani ngalah luhur wekasane, tumungkulo yen dipun dukani, bapang den simpangi, ana catur mungkur, mbangun turut*". It means that in order for us to be useful or successful people, we must behave politely to people. Otherwise, not being arrogant, placing others higher, respecting others in relationships, daring to give in and yielding does not mean losing, being able to control oneself, not fighting back, looking down when wrong and being scolded, accepting input from others, avoiding things that not good, not extravagant, simple, and avoiding fighting, gossip, and prejudice are also necessary.

These values are the identity of the Mataraman community, manifesting themselves as true human beings. Besides that, it is also a reference in realizing an ideal life as a Javanese man who is physically and mentally prosperous. The virtues possessed by the Mataraman society above are in line with Javanese manners as revealed by the Yogyakarta Province Character Development Team as follows: (1) dedication; (2) honesty; (3) courtesy; (3) tolerance; (4) discipline; (5) sincerity; (6) responsibility; (7) harmony; (8) politeness (*tepa selira*); (9) self-awareness (*papan, empan, adepan*); (10) manners (*tata krama*); and (12) mutual cooperation (*gotong royong*) (Supriyoko 1996).

Meanwhile, according to Endraswara (2016) the essence of Javanese culture is "Javanese people believe in and take refuge in the God, the cause of all life, the existence of the world and the whole universe, the beginning and the end; Javanese people are harmonious in peace, harmonious (*rukun*), orderly in their birth and peaceful in their mind, noble and human beings".

Based on the explanation above, virtues of the Mataraman community are: (1) simplicity, (2) respect for others and parents, (3) politeness, (4) frugality and caution (*gemi, nastiti, ati-ati*), (5) harmony, (6) surrender to God Almighty, (7) prudence, wisdom, and (8) humbleness.

The virtues of Javanese culture developed by the Mataraman community are no different from the focus of developing the character of the Indonesian nation which lies in six values, namely three personal dimensions (honesty, common sense, and courage), and the other three social dimensions (fair, responsible, and tolerant). If the development of these six values is truly developed in a professional and national scope, it is not impossible that an increase in the quality of Indonesian people within one generation will occur (BSNP, 2010). In terms of instilling the values of Javanese character values, Javanese character includes *taqwa*, remembering of God, *tawakal*, repentance, shame, fairness, respect for others, sincere, patient, honest, kind, forgiving, helpful, grateful, wise, benevolent, brave, officer and loyal, then the tim pengembang budi pekerti (TPBP) province DIY stated Javanese character includes: devotion, honesty, courtesy, tolerance, discipline, sincerity, responsibility, friendship, *tepasaliro*, small board, manners, and *gotong royong* (Supriyoko; 2000).

Another opinion about the character that needs to be developed in Indonesian society was expressed by Megawangi, according to him, there were nine noble attitudes instilled from an early age in Indonesian children at school. These attitudes are: (1) love of God and truth, (2) responsibility, discipline and independence, (3) trustworthiness, (4) respect and courtesy, (5) compassion, care, and cooperation, (6) trust self, creative, and never give up, (7) justice and leadership, (8) kind and humble, and (9) tolerance and love peace. Then, other characters that need to be developed are "productive" and "creative/innovative" characters in thinking and working (Megawangi, 2004).

3.2 *The actualization of the Mataraman society virtues*

Actualization means the manifestation or showing of ownership of something. In the context of this study, actualization is showing the ownership of values in the form of Mataram society behavior, such as speech, attitudes, appearance, and actions.

Thus, the actualization of the virtues of the Mataraman community in their daily lives is manifested when they interact with each other through communication, appearance, attitudes and actions while in the family, community and during educational activities and at work. This is in accordance with the opinion of Spencer-Oatey (2004) that society's culture reflects the values of politeness and character as cultural products that are realized in the way of speaking and behaving.

The actualization of Javanese cultural virtues in the above manner has been conveyed by Gertz (1985) that Javanese children learn the principle of respect through 3 situations, namely wedi, isin, and shy. Wedi or fear, most Javanese children carry out the orders and prohibitions of their parents at first with fear. Isin or is called the culture of shame, shy is defined as doing everything that is not with sincerity, doing it only because of feeling bad for those who rule. The practice of character education by cultivating a sense of wedi, isin, and reluctance is very effective in developing children's character to behave in accordance with the noble Javanese cultural values.

Simple value actualization in the Mataraman community is manifested in the form of the appearance of the house where they live, the characteristics of the Mataraman community house in front are very simple but very luxurious inside. The mataraman family house tends to appear closed when viewed from the front, the house materials are very good, the door is in the style of butterfly fighting (*model kupu tarung*), the cleanliness of the house is well maintained and always maintained. Likewise, in dress, as said by Mrs. Sujilah as "*Ndhuk yen nggawe rok aja cingkrang cingkrang, saru lho, lan ora usah macak sing aneh aneh, prasaja ae.*"

The actualization of the value of respect for others, especially parents, according to the statement of Mr. Willis Purwanto (a relative of Mataraman) that "*Kagem tatanan jawi kurmat kaliyan tiyang sepuh puniko sanget utami, awit sanget mboten mapan pantes menawi ningali wonten lare-lare nungkak krama kalihan tiyang sepuh, menapa malih menawi ngantos mboten gadhah wedi kalihan tiyang sepuh, mboten duwe isin yen gadhah kalepatan, mboten sungkan nglampahi ingkang kirang sae, waah kedadosanipun nggih dados lare kurang ajar.*"

The actualization of the value of respect for parents is most evident in the tradition of *sungkeman.* Tends to Mrs. Sujilah As follows, "*Sungkeman menika ajeg saben ba'da utaminipun, nanging sampun mboten kados rumiyin. Menawi kulo rumiyin taksih laku dhodhok, sakmeniko sampun mboten.*" The custom in the Mataraman family is that every family gathers, in turn, from the old to the young, to hold prayers, to forgive each other and most importantly to ask for prayers from the mother and father.

In addition, in terms of decision making, in various areas of life, the Mataraman community has great respect for parents, as said by Mrs. Purnawati (50 years old) as follows, "*Yen babagan sing gegayutan karo keputusan kuwi bapakmu le, mengko ibu bisa keno duko.*" The child's trust in the father is very different from that of the mother. Children are more afraid and obedient to the father. What the father said, usually followed by the child and wife. There is a saying "*Yen ora manut tak aturke bapakmu lo.*" The tendency in reciprocal interaction between the child and the mother uses the Javanese language *ngoko*, while with the father uses the *kromo* language.

Actualization of God values (*ketuhanan*), trust and devotion to the God through the tradition of salvation, *ngampirne neton*, by means of *njenang abang*, *buceng kuat, brokohan.* As said by

Mr. Setyo Winarno: *"Slametan meniko sae awit budaya, nanging kulo tansah nyuwun dhateng masyarakat sampun ngantos ukawis-ukawis donga mboten nerak wewaler agami, umpami menawi rumiyin nyenyuwunipun dumateng cikal bakal utawi dhanyangipun, nanging sakmeniko nyenyuwunipun tumuju dhateng Gusti Ingkang Maha Kuwaos dene para leluhur ingkang sampun sumare dipun suwunaken pangapuntun supados anggenipun kondur dhateng Pangayunaning Gusti saged diapuro sedaya dosanipun, kaparingan papan ingkang sekeco."* That is what Mr. Sutadi (a resident of Mataraman) conveyed, as follows "*Wonten tlatah kulo meniko taksih nguri nguri budaya, tradisi ugi taksih dipunlenggahi, biasanipun amargi meniko tradisi slametan kajatan meniko taksih glenggahi, slametan taksih wonten, sajen kajatan slametan taksih wonte.*"

The basic character that becomes the foundation of the Javanese is the *nrimo* attitude (Endraswara, 2016). *Nrimo* is accepting everything with spiritual-psychological awareness, without feeling *ngrundel* (grumbling). Whatever is accepted sincerely (*paringane sing kuwasa*). For self-control, as said by Mr. Sutadi, when giving his son advice when choosing one of the jobs, "*Wis to le ditampa ae, nrima paringane sing kuwasa, aja nganja anja sing sumeleh.*" This attitude shows the guidance of patience associated with the *garis pepesthen yang maha kuasa*.

The cultural values of introspection, *mawas diri, nandhing sariro, ngukur sariro, tepa saliro, mulat saliro,* are manifested in associations with fellow humans starting from the family, an attitude of *rasa rumangsa, tepa salira* are instilled, which is often done in symbol language, under certain circumstances. it is enough to use the symbol to put the index finger on the mother's mouth. For example, when the father is sleeping, the mother puts her forefinger in her mouth while saying, *"Huss aja gojek, bapak sare, mundhak kebribenan."*

The manifestation of the values of *gemi, nastiti, ati-ati, arif, dan rendah hati* is done when doing something must be calculating and careful, not careless. This statement is in accordance with the narrative of Mr. Setyo Winarno (a resident of Mataraman Village) *"menawi tiyang gesang meniko kedhah geni lan ati-ati supados slamet lan tenterem uripe, kaliyan sesame inggih kedah rendah hati".*

The manifestation of the value of harmony is carried out by giving expressions such as those conveyed to others such as "To achieve success one must be harmonious;" *crah agawe bubrah*; *rukun agawe santosa*, and so on. The actualization of the virtues of the Mataraman community is manifested through communication using the right language, as conveyed by Mrs. Endah, "Actualization of Javanese language and *trapsila* there are two things that cannot be separated between *Pocap and Patrap* must be harmonious and balanced." *Pocap* is a choice of words that are spoken verbally and *Patrap* is a balance of past behavior when saying a word. For example, when the child says the word "*Inggihm,*" it must be accompanied by the head bowing, the body half bent, hands *ngapurancang*.

The actualization of the virtues of the Mataraman society is also manifested in the form of various expressions, which are used as guidelines for people's behavior, to live side by side in harmony, harmony and balance between the life of the world and the hereafter. Balancing life in harmony with the environment or conditions that are happening. Besides that, it can also be used as a control to behave well in human relations with fellow humans, humans with the environment, and humans with God.

Based on the explanation above, the actualization of the virtue of the Mataraman community in everyday life is manifested in the form of patterns of use of language and attitudes when communicating with others, personal appearance and place of residence, attitudes and actions when interacting with others in various activities both during at home, in the community and at work.

4 CONCLUSIONS

There were eight virtues of the Mataraman society. They were simple, respect for others and parents, introspection or selfawareness, frugal and careful, harmonious, surrender to God Almighty, wise, and humble. Actualization of the virtues in life with language, use of words, volume and intonation, behavior, clothing, leadership in the family, and traditions. Based on the research findings, be suggestions for making the virtues as a guideline in behavior and behaving in accordance with the virtues.

REFERENCES

BSNP. 2010. *Paradigma pendidikan nasional abad XXI.* Badan Standar Nasional Pendidikan. Jakarta. Depdiknas.

Endraswara, S. 2016. *Budi pekerti Jawa*. Cakrawala. Jogjakarta.

Geertz, H. 1985. *Keluarga Jawa*. Gratifi Pers. Jakarta.

Megawangi, Ratna. 2004. *Pendidikan karakter: Solusi yang tepat untuk membangun bangsa.* Cimanggis: Indonesia Heritage Foundation.

Moedjanto. G. 1987. *Konsep kekuasaan Jawa*. Kanisius. Jogjakarta.

Supriyoko, K. 1996. *Budi pekerti sebagai syarat kelulusan*. Jogjakarta: 19 Mei 1996.

Spencer, H. & Oatey.T. 2004. *Culturally speaking: Managing rapport through talk across culture.*

Empowering Civil Society in the Industrial Revolution 4.0 – Sukowiyono et al (Eds)

The influence of tourist village status on the perception of happiness among residents

S.D.E. Putra*
Universitas Negeri Malang, Malang, Indonesia

K. Rahman & N.A. Fiaji
Brawijaya University, Malang, Indonesia

ABSTRACT: This study examined the local community's happiness perceptions in Kampung Wisata Jodipan, an urban tourist village in Malang since it was changed from a previously slum area. It is famous as Kampung Wisata Warna-Warni Jodipan, a colorful urban tourist destination since houses are painted in many colors. This study described the most influential variables determining happiness and changes in the local community's perceptions of happiness after the area was changed into a tourist destination. The study employed a quantitative descriptive approach using questionnaires with indicators measuring happiness perceptions: time management, life balance, oral culture, religiosity, mental and physical health, and solidarity. Direct interviews validated data collected using questionnaires. Results confirm that compared to its previous status, the community now has better (1) oral culture and religiosity, (2) mental and physical health, and (3) solidarity, togetherness, and caring for others.

Keywords: urban village tourist destination, happiness, Kampung Jodipan

1 INTRODUCTION

Tourism is one of the most important elements in national economic development. In accordance with the Ministry of Tourism report (2016), the country's foreign exchange income in the tourism sector has increased from IDR 144 trillion in 2015 to IDR 176-184 trillion (prognosis) in 2016. The main source of increasing foreign exchange income in the tourism sector is the increasing number of tourists, from 10.4 million in 2015 to 12.02 million in 2016. Thus, it is not surprising that the government launched tourism as one of the core economies in all regions of Indonesia with great potential in the tourism sector, including the development of tourism that is currently a hype in Malang.

Settlements in Malang city, where the topography is in the form of hills and river slopes, make the people of Malang more creative in building houses. Many housing estates were built with attractive designs by both elite housing developers and local communities. However, many housing or settlements that are owned by local residents of Malang City were established in an irregular manner and even tended to be slum. Some of the slums can be easily found in Malang, such as in the land belonging to Railway Bureau Company (PJKA), Jodipan village, Comboran area, settlements in the village of Mergosono, residential areas of the Meror, and other places that are still in the city of Malang. Specifically, the Jodipan Subdistrict, a slum area under the Blimbing sub-district bridge in Malang, has transformed from a slum settlement into a tourist attraction in Malang City as the new colourful village of Jodipan.

The "colourful" village of Jodipan is one of the villages on the banks of the Brantas Watershed which is now the object of tourism in Malang City. The village that is currently viral among

*Corresponding Author

DOI 10.1201/9781003180128-22

these tourists, was formerly a slum settlement which included the planned eviction of the Malang City Government. However, after being polished and beautified by Guys Pro community (from Muhammadiyah University Malang students) and mural artists in Malang City, Jodipan village turned into one of the mainstay tourism destinations in Malang.

Change in status from slum villages into tourist villages greatly affects the economic growth of the community. This is a significant development for the city of Malang. With the addition of new tourist spots that can attract many tourists for more along with visiting Malang, it means opening up new economic opportunities. Moreover, the city government of Malang is currently seriously activating the tourism sector to boost the economy of the community, apart from industry and education, as explained in Tri Bina Cita Malang.

The development of a good tourist village cannot run positively if the community does not participate in encouraging its development. These changes must have greatly influenced the lives of the people around, especially changes in economic resources, which then had an impact on the social psychology of the community itself, especially in providing perceptions of happiness. However, is economic progress itself the only source of happiness for the surrounding community, especially for the Jodipan Village community? According to Carr (2004) happiness is a positive psychological condition, which is characterized by high satisfaction with the past, high levels of positive emotions and low levels of negative emotions. Tourism as a phenomenon of human life has never been free from debate and analysis. This research is intended to fill in the information gap on how tourism as an economic, social, and cultural event is influenced by a variety of complex internal and external factors. Foresight in understanding the character of tourism and its external environment undoubtedly makes it easier for decision makers to use it as a nation-building tool, and for the community to motivate the development of community empowerment.

In the tourism bureaucracy, it is echoed as a "panacea" that is able to overcome poverty into material prosperity. The contribution of the tourism sector to the acquisition of foreign exchange and the creation of employment on a macro level is quite significant. Reports of various expert studies concluded that tourism contributions that significantly affect the economic development of a country or region appear in the form of expanding employment opportunities, increasing income (foreign exchange) and equitable spatial development (Damanik, 2017). So that the changing of Jodipan village into a tourist village will change people's lives in private, such as patterns of conflict within the family that should be avoided when tourism hours are taking place, then the level of privacy can no longer be owned by the Jodipan community because of the branding of tourist village inevitably the people have to accept happily the many tourists who visit the village regardless of the time. An account of Jodipan village and its community life and all kinds of problems that might arise in it indirectly can be a topic for research that can increase the repertoire of knowledge in the realm of social life. Thus, this study is carried out with the title "The Effect of Tourist Village Status on Perceptions of Community Happiness in Jodipan Village."

2 METHODS

This study intends to determine the level of relationship between variables in a population so that this study can be called the correlational study. This is based on the characteristics of connecting two or more variables, the magnitude of the relationship is based on the correlation coefficient, in view of the relationship manipulation is not done and the data is quantitative (Zuhrial, 2007). This study not only explains but also ensures the magnitude of the relationship between variables. The relationship between these variables is an asymmetrical relationship where the variables in the study change simultaneously (Ghony, 2009). In other words, changes in the independent variable are also followed by changes in the dependent variable.

The determination of the sample in this study used the survey method, because the number of family heads is fewer than 100 people. We determined the main sample of 50 people in the colourful village of Jodipan and 50 people in the village of Jodipan who were not colourful, hoping to find out the gap or difference between the Jodipan village which is a tourist status and that is not a tourist status.

This research design used Human Centered Design in the process of creating something based on common natural characteristics and focuses heavily on the psychological state and human perception. Whatever is designed is human-centered based on psychology, physiology, sociology, and other sciences that analyse human life and its interactions with the environment. In this study, using a user experience questionnaire that has been conducted on 50-100 respondents, it was collected and calculated using SPSS 16 tools.

This study attempts to examine the relationship between independent variables (X), namely the status of tourism villages with the dependent variable (Y), which is the perception of happiness. Furthermore, this study also examined the relationship between attitude sub variables with sub variables of perception of happiness. The perception of happiness that is the measure in this study is based on An Extensive Analysis of GNH Index of the Centre for Bhutan Studies, 2012. Variables used to measure perceptions of happiness include psychological wellbeing, health, times use, education, cultural diversity and resilience, good governance, community vitality, ecological diversity, and living standards (Ura, 2012). The measure was also used by the Central Statistics Agency to be reduced to more complex variables. Therefore, in this study the measurement of the perception of happiness is limited to four variables, namely: utilization of time and life balance, speech culture and religious values, physical and mental health, and solidarity among community members. The form of relationship model framework in this research is as follows: X: Tourist Village Status, Y: perception of happiness, Y: Utilization of Time, Y2: speech culture and religious values, Y3: physical and mental health, Y4: solidarity among community members.

3 RESULTS AND DISCUSSION

The results of this study were discussed by analyzing the results of the variables with a theory that explains the ideal place of tourism and ideally the happiness achieved by the community in the tourist area.

3.1 *Analysis of variable correlation result with tourism theory*

As ideal tourism is written in accordance with Law No. 10 of 2009 must be based on benefits, kinship, fairness and equality, balance, independence, sustainability, participatory, sustainable, democratic, equality and unity, all of which are realized through the implementation of tourism development plans by taking into account the diversity, uniqueness and distinctiveness of culture and nature and human needs for travel. When juxtaposed with the results of data collection in the Jodipan Tourism Village, an alignment or contradiction will be obtained as follows:

a. The principle of benefit and continuity of tourism status does indeed have harmony with the theory because of the fact that the community is very happy with the change in tourism status in Jodipan Village. Especially in the use of time where the community is more disciplined, orderly to organize the village so that many tourists arrive and fulfillment of their food is better because of the increased income for the restoration of the environment. In the aspect of physical and mental health there are also statements that support that tourist are included as part of the source of income.
b. The principle of kinship, unity and participation is mostly in harmony with the results of the study, but there are some items that show inconsistent statements that occur in Jodipan Village. Harmony is characterized by always prioritizing the public interest, caring for others, easy to share, be compact, fair and advise each other on the right path. Whereas the inconsistency is a little fanatical even though the group is wrong to be defended, and it has been slightly contaminated by the individualist nature because there are three statements that describe individuality, two of which have high values, while one statement shows a low fanaticism level.

c. The principle of justice, balance and democracy is very harmonious with the Jodipan community, this is explained in the aspect of community solidarity. Likewise, the aspects of speech culture and religious values where the relationship is very strongly related to happiness and freedom of expression.
d. The principle of independence in terms of building their aspirations is able to exchange ideas and ideas well, because they have flexibility and friendliness in communication. However, some people have not been able to gain the independence in realizing both physical and mental health because there is a statement where one has not been able to manage stress, and awareness of maintaining physical fitness is still low
e. The principle of diversity and uniqueness is still not reflected optimally because the unique colorful concepts are also derived from concepts that already exist in other countries.
f. The principle of environmental sustainability has been partly realized in Jodipan village in terms of cleanliness of riverbanks and settlements. However, hygiene awareness is still forced by rules because some residents still have problems with food sufficiency.

3.2 *Happiness analysis achieved by Jodipan tourism village*

The analysis of happiness in Jodipan Tourist village will be seen if it is collided with Jodipan Village which is a non-tourist village. The extent of happiness achieved by Jodipan Village which has a tourist village status will be measured.

To find the gap, the researchers conducted questionnaires as well as samples of the Jodipan community outside the tourist village status to find out the independent comparison of the sample. The results of Pearson's analysis about the happiness residents of Jodipan's Village can be seen at table 1.

Table 1. The results of Pearson's analysis using the SPSS 16 application, the happiness in the aspect of time utilization tended to be significant in Jodipan's colorful tourist village

Indicators item	Reliability Statistics
Time Utility (variable Y1)	0.693
Speech Culture and Religious Values (variable Y2)	0.732
Physical and Mental Health (variable Y3)	0.716
Solidarity among Community Members	0.712

The results of the significance of each variable are obtained as follows:

a. Time utility (variable Y1)
There is no significance either in the status of Jodipan tourist village or not a tourist village status in Jodipan, with a score of 0.624, which means there is no significant difference in happiness in time utilization among the community in terms of the tourist village status of Jodipan Village.
b. Speech culture and religious values (variable Y2)
There is a significant difference between the parts of the village with tourist village status in Jodipan with those that are not, with a score of 0.04, which means that the Jodipan community living in the tourist village area is happier in aspects of speech culture and religious values.
c. Physical and mental health (variable Y3)
There is a significant difference between Jodipan village which has a tourist status and a non-tourist status in the aspect of physical and mental health with a score of 0.007, meaning that the Jodipan community in the tourist village is far healthier.
d. Solidarity among community members (variable Y4)

There is also a significance that the tourism status of Jodipan village greatly influences the community solidarity, with a score of 0.031, which means that the Jodipan community with tourism status is more solid, caring, and friendly with each other.

After conducting direct verification through interviews with residents of Jodipan, research on changes in Jodipan's perceptions of society is in accordance with the initial hypothesis. Many community members responded positively to changes in the status of the Jodipan village, from a village that was considered as a slum to become a tourist village. The change is seen from changes in people's income that is higher than before.

4 CONCLUSIONS

It can be concluded that there is significant influence of Jodipan's colourful tourist village status on the happiness of its people, even though not in all aspects. Detailed aspects affecting the happiness of the Jodipan community are as follows: (a) the Jodipan community expresses happiness in a speech culture and religious values is far better than a village that is not a tourist village, (b) the Jodipan community has far better physical and mental health conditions when the village has tourism status, c. solidarity, cohesiveness and caring for each other among Jodipan community members is better as a tourist village.

REFERENCES

Carr. A. 2004. Positive psychology, the science of happiness and human strength. New York: Brunner-Routledge

Damanik, J. 2017. Pariwisata Indonesia antara peluang dan tantangan. Yogyakarta: Pustaka Pelajar.

Ghony, D. 2009. Metode penelitian pendidikan pendekatan kuantitatif. Malang: UIN Malang Press

Humas. 2016. Praktikum mahasiswa komunikasi UMM lahirkan Kampung Warna-Warni Jodipan. Retrieved from: http://www.umm.ac.id/id/berita/praktikum-mahasiswa-komunikasi-umm-lahirkan-kampung-warna warni-jodipan.html).

Ministry of Tourism Report. 2016. *Laporan akuntabilitas kinerja Kementerian Pariwisata.* Retrieved from: http://www.kemenpar.go.id.

Sunaryo, B. 2013. Kebijakan pembangunan destinasi pariwisata: Konsep dan aplikasinya di Indonesia. Yogyakarta: Gava Media

Ura, K, et al. 2012. An extensive analysis of gnh index. Thimphu Bhutan: The Centre for Bhutan Studies.

Yuwanto, E. 2017. Wawali Kota Malang pariwisata berperan dongkrak ekonomi. Retrieved from: http://nasional.republika.co.id/berita/nasional/daerah/17/10/02/ox69e2438-wawali-kota-malang-pariwisata-berp eran-dongkrak-ekonomi).

Zuriah, N. 2007. Metodologi penelitian sosial dan pendidikan. Jakarta: Bumi Aksara.

Empowering Civil Society in the Industrial Revolution 4.0 – Sukowiyono et al (Eds)

The values of local wisdom contained in the Grebeg Sabrang traditional dance at Mangun Dharma art gallery

F.C. Chanada*, D. Sukriono & R. Umar
State University of Malang, Malang, Indonesia

E.N. Hapsari
Universiti Malaya, Kuala Lumpur, Malaysia

ABSTRACT: This study discussed the values of local wisdom contained in a traditional dance called Grebeg Sabrang dance, precisely at Padepokan Seni Mangun Dharma. This study aimed to introduce one of traditional dance that is still original from Padepokan Seni Mangun Dharma and the benefits provided by the Mangun Dharma Art Gallery to the Grebeg Sabrang dance in the preservation of local wisdom values. This study used a qualitative approach with descriptive study. Data collections techniques used interviews, observations, and documentation. Data analysis technique used interactif analysis. The results of this study showed Grebeg Sabrang Dance is a Malang traditional dance which is still original and has a lot of values of local wisdom.

Keywords: Grebeg Sabrang, traditional dance, local wisdom values

1 INTRODUCTION

Indonesia is a country that has a wide variety of cultures, ethnicities, races, religions and customs. Each region has its own characteristics that are unique and differentiate between regions, because the culture referred to here is an art that can be developed independently by the surrounding community. People are free to work and create according to their respective regions. Each region has traditional rituals, traditional ceremonies, and traditional dances that are different and unique to the area.

Dance is the rhythmic and beautiful movements of the human body. Every dance has its own beauty, and that beauty is relative. Dance is a combination of several elements, namely *wiraga*, *wirama*, and *wirasa*. The movements created in a dance must be in harmony with the accompanying music and there is an inspiration when dancing a dance (Sumaryono in Rahayuningtyas, 2015).

Besides the element of beauty, in a dance there are also values. In dance, there are symbols of life that have deep meaning and values about the nature of life (Pratiwi, 2016). In a traditional dance, the movements that are created usually contain symbols or improvements from the movements carried out, and are characteristic of an area. Based on cultural statistical data Hadi et al., 2019) Indonesian arts established by UNESCO in 2018 are Balinese dance, Saman dance, and Wayang performances.

Grebeg Sabrang dance is one of the traditional dances that must be maintained, because this is a traditional dance from the Malang area which is characteristic of the Malang area. *Grebeg Sabrang* dance is already attached to the Malang regional identity because the dance movement is a "*Malangan*" movement and is a standard dance that must always be maintained. There are local wisdom values contained in the *Grebeg Sabrang* dance. The importance of preserving the *Grebeg*

*Corresponding Author

DOI 10.1201/9781003180128-23

Sabrang dance is because this dance is a standard dance which until now has been made an icon of the Malang area.

It is very important to provide knowledge about traditional arts to the community, so that the values contained in a traditional art can be conveyed. The values contained in an art or culture will not emerge by itself if it does not try to express it in line with the many foreign cultures that have entered Indonesia (Gunawan & Sulistyoningrum, 2013).

2 METHODS

This study used a qualitative approach with descriptive study. Descriptive research according to Creswell is a research method that seeks to describe and interpret objects as they are (Sangadji and Sopiah, 2010). This study was conducted by the author by being present in the field, precisely at the Mangun Dharma Art Gallery, Tumpang Sub District, Malang Regency. The data collections techniques used interviews, observations, and documentation. The data analysis technique used is Cresswell's interactive model which includes organizing and preparing data, reading all the data, making coding, using coding to describe, connecting between themes, and providing interpretation (Sugiyono, 2018).

3 RESULTS AND DISCUSSION

3.1 *The origins of the Grebeg Sabrang dance at the Mangun Dharma art gallery*

Grebeg Sabrang dance at Mangun Dharma Art Gallery is one of the traditional mask dances typical of Malang area created by an artist named Ki Soleh Adi Pramono in 1983When it was first created; the dance was called *Grebeg Brang Wetanan* Dance. In 1989, he officially founded an art studio called *Padepokan Seni Mangun Dharma* (Mangun Dharma Art Gallery). The *padepokan* houses various kinds of arts.

In 2001, the *Grebeg Sabrang* Dance was created by him to represent the Malang region as for the Indonesian Mask Dance event. Grebeg Sabrang dance is called the *Grebeg Bolo Werto* dance which means that *werto* is red. Then the name became the *Grebeg Sabrang* Dance which until now is known to the community around the Mangun Dharma Art Gallery. This is in accordance with the opinion of Sumaryono (2011) as quoted by Rahayuningtyas (2015) that dance is rhythmic and beautiful human body movements. The beauty of the movement can be seen from the non-physical side, namely the impression that comes from the dance arrangement. *Grebeg Sabrang* dance is a manifestation of a dance stylist or choreographer at the beginning of creating the dance and becoming a representative for the Malang area.

3.2 *Characteristic of the Grebeg Sabrang dance at the Mangun Dharma art gallery*

3.2.1 *Variety of movements*

The characteristics of the variety of movements are still original and have distinctive features. This can be seen from the addition of movement at each change of music. In addition, it has a large volume. The various movements include *kanon* movements, *balan* movements, there are top, bottom, middle designs, and so on. While the details of the dance movement include *singget*, *gedheg mantuk*, *labas*, *trecet*, *gedrug* or *gedrugan*, and *ronggo lawe*. The dashing character in the dance is emphasized by the hand attitude, namely the closed *supit* and the leg attitude which is called *tanjek*. The development movement includes *maju gawang* and *solah* motion.

3.2.2 *Gendhing*

The *Grebeg Sabrang* Dance at Mangun Dharma Art Gallery has a piece whose dominance is *ayak-ayakan* or so called *krucilan* accompaniment, because in the past it was used to accompany *wayang krucilan*. A gendhing is determined by *cengkal kendang*. *Gendhing* for the *Grebeg Sabrang* Dance at Mangun Dharma Art Gallery is 5 panthers, namely there are 10 pathets whose dominance is 4 pelog rhythms, because this dance is an all-night dance. Then the line is pathet 8, in pathet 8 there is *maujengkaran* then go up to pathet 9 and return to pathet 8 again. After there is the first *jejer*, then there is the second *jejer* which is followed by Gunung Sari Dance.

3.2.3 *Costume*

The costumes from the *Grebeg Sabrang* Dance at the Mangun Dharma Art Gallery include: (1) *Grebeg Sabrang* masks; (2) *Jamang*; (3) Wigs; (4) Neck decoration; (5) Shoulder hanger; (6) Bracelet; (7) Pants; (8) *Jarik*; (9) Thigh Cover; (10) Black *stagen*; (11) Belt; (12) *Sampur* or shawl; (13) *Keris*; and (14) *Gongseng*. For the masks in the *Grebeg Sabrang* Dance, there are five characters consisting of: (1) Patih Lindumanon; (2) Patih Lindusekti; (3) Kala Memreng; (4) when wrapping his face; and (5) Gajah Sugwa.

3.2.4 *Staging*

There are two parts or episodes in the performance. In the first part the dancer enters the stage to dance the first part. In the second part, dancers enter the stage to describe the story of the soldier on the opposite side with the atmosphere of a journey that has approached the battlefield or the place to carry out their duties. The appearance or performance depends on the storyline that is performed by the *dhalang*.

3.3 *The meaning of Grebeg Sabrang dance at the Mangun Dharma art gallery*

3.3.1 *Social anthropology*

Grebeg Sabrang dance at *Padepokan* Seni Mangun Dharma means procession depicting soldiers coming out of the *Alun-alun* (the front yard of a kingdom). Patricia Dewar in the Journal of Dance Ethnology quoted by Rahayuningtyas (2015) states that dance as a movement communication system has social functions and uses. The meaning of the dance is a soldier from Prabu Klana Sewandana who is carrying out state duties. The large number of troops in the royal square makes it called *grebegan*. From a social perspective, this dance is seen from the meaning of the storyline and the elements contained. The narrative referred to is that the soldiers are carrying out state duties together. While the elements of the *Grebeg Sabrang* cannot stand alone. There are *gamelan* accompanists, costume stylists and spectators or audience.

3.3.2 *Art anthropology*

The performance is in the form of *proseniumtis,* which is a frame stage with the audience in one direction. The law of art in *Grebeg Sabrang* Dance is a choreographer who pays attention to group designs. The group dance design has a variety of elements which include the density of the number of dancers, floor patterns, and *gendhing*. The law of choreography is divided into two, namely meaningful motion and pure motion. Meaningful motion means having meaning or meaning in dance. For example, a movement called *ronggo lawe* in this dance depicts a king examining his army or soldiers. While pure motion has no meaning and is an original movement.

This is in line with the opinion of Rahayuningtyas (2015) that dance movements are movements that have undergone a change or stylization process from wantah and meaningful movements. The motion of *wantah* is very easy to understand because it is an original motion that has not yet received a touch of beauty. Meanwhile, the meaningful motion is a beautiful motion and has a lot of meaning.

3.3.3 *Philosophy anthropology*

Philosophical anthropology includes two physical and four inner keys. The two keys of birth or choreography include *kawruh mengku* and *kawruh megeng*. The inner keys include *mapak*, *mengkal*, *megeng*, and *mapan*. Each key has its own meaning. *Mapak* means pick up the rhythm, in which a dancer dances first to provoke *gendhing*. *Mengkal* is in accordance with the character who is detached. *Megeng* is an inspiration in a dance. *Mapan* is the harmony between the various movements and *gendhing*. If all keys can be mastered by the dancer, then a dancer can animate or unite with a dance.

3.4 *Local wisdom values in Grebeg Sabrang dance at the Mangun Dharma art gallery*

According to Mulyana, quoted by Frimayanti (2017) that value is a reference to belief in determining an option. Local wisdom according to Suyatno (2011), can be defined as a local cultural wealth that contains a life policy, namely a way of life which accommodates life wisdom. Almost every local culture in the archipelago is known for its local wisdom that teaches mutual cooperation, tolerance, work ethic, and so on. On the other hand, local wisdom can also be said to be a view of community life in the form of activities and used as a guide in carrying out various activities (Astuti, 2016).

3.4.1 *Religious value*

Local wisdom aims to increase welfare and create peace. The religious value is contained in the *Grebeg Sabrang* Dance. The religious value is shown in the appearance or performance and making of masks. In ancient times, people danced for good and approached God by means of art. Art is not just a thematic dance, but art is related to mystical science. The *Grebeg Sabrang* dance at the Mangun Dharma Art Gallery is associated with mystical science, both in the appearance and in the process of making masks.

3.4.2 *Mutual cooperation value*

The general meaning contained in the values of mutual cooperation depends on the form of mutual cooperation that is carried out, because each mutual cooperation has different values and meanings (Rolitia et al, 2016). The value of mutual cooperation shown in the storyline of the dance. The storyline in question is a depiction of the *Grebeg Sabrang* Dance. It depicts a group of warriors from Prabu Klana Sewandana who are fighting to carry out state duties.

3.4.3 *Moral value*

Moral values are educational values, namely educational values which include individual attitudes in personal life, social life, and life related to God (Nugraheni, 2018). The moral value is shown in every tradition that is still maintained today. The majority of the traditions are still maintained today, both from making masks, various movements, *gendhing*, and costumes. This is because Ki Soleh Adi Pramono respects the services of his ancestors, by maintaining traditions and customs. The making of masks in this dance is still very sacred and upholds tradition. This is evident from the highly respected politeness and ethics. Subadio (1986) as quoted by Brata, that local wisdom (local genius) as a whole includes, maybe even be considered the same as cultural identity which can be interpreted as the identity or cultural personality of a nation.

3.4.4 *Togetherness value*

The value of togetherness is shown in the appearance or performance of the dance. Performances always prioritize togetherness, so that the choreography and pieces can run in harmony. In addition, so that the meaning contained in the *Grebeg Sabrang* Dance can be conveyed.

3.4.5 *Educational value*

Values and education are aligned with a process of determining good and right in the framework of forming a person's personality towards a more mature direction and always adapting to their environment (Widagdo, 2014). The *Grebeg Sabrang* dance at Mangun Dharma Art Gallery contains educational values shown in the embodiment of the dance. Apart from the Mangun Dharma Art Gallery, this dance is also taught at schools and foreign dancers. *Grebeg Sabrang* dance is a traditional dance intended for the general public, so that everyone can learn the dance directly at the hermitage or gallery. Therefore, Mangun Dharma Art Gallery provides facilities for those who want to learn the dance. This is in line with the opinion of Santosa (2017) that local cultural wisdom is local knowledge that has been so integrated with belief systems, norms, culture, and expressed in traditions and myths that have been followed for a long time. The focus of the function and role of local wisdom is on three aspects, namely local wisdom, character, and character education. These three aspects can certainly be explored as cultural elements that are believed to be a strong binding force for the integrity of the Republic of Indonesia.

3.4.6 *Art value*

Cultural identity markers, for example, can come from a uniqueness that is believed to exist in religion, language and customs in the culture concerned (Maunati, 2004 in Brata). The art value is shown in the traditions that are in it, because the dance is still original and still maintains the tradition until now. *Grebeg Sabrang* Dance at Mangun Dharma Art Gallery is a unique traditional Malang art. Apart from being the identity and "icon" of the Malang area, the uniqueness lies in the characteristics of this *Grebeg Sabrang* Dance and does not exist in other arts.

3.5 *Benefits given by Mangun Dharma art gallery to the Grebeg Sabrang dance in preserving local wisdom values*

As a cultural heir, Mangun Dharma Art Gallery strives to maintain traditional arts which are characteristic of the Malang area. Both from the variety of movements, costumes, *gendhing*, meanings, and so on. It can be seen from the results of all the choreography and music creation processes of the *Grebeg Sabrang* Dance which have survived until now and there are still enthusiasts who want to learn this dance.

4 CONCLUSIONS

The origins of the *Grebeg Sabrang* Dance began in 1983 when Ki Soleh Adi Pramono would present a dance that represented the Malang area. The characteristics include a variety of movements, *gendhing*, costumes, and staging or performances. The meaning of the *Grebeg Sabrang* Dance includes: (a) Social anthropology; (b) Art anthropology; (c) Philosophical anthropology. The values of local wisdom in the traditional *Grebeg Sabrang* dance at Mangun Dharma Art Gallery can be concluded that there are six values of local wisdom, namely: (1) Religious value; (2) Mutual cooperation value; (3) Moral value; (4) Togetherness value; (5) Educational value; (6) Art value. Mangun Dharma Art Gallery provides defense efforts to all elements and traditions of the *Grebeg Sabrang* Dance.

REFERENCES

Astuti, F. 2016. Menumbuhkan nilai-nilai kearifan lokal melalui karya tari koreografer perempuan di Sumatera Barat: Suatu tinjauan gender. *Humanus: Journal of Humaniora Sciences* 15(1): 53–60.

Brata, I.B. 2016. Kearifan budaya lokal perekat identitas bangsa. *Bakti Saraswati Journal* 5(1): 120–136.

Frimayanti, A.I. 2017. Implementasi pendidikan nilai dalam pendidikan agama Islam. *Islamic Education Journal of Lampung University* 8(2): 227–244.

Gunawan, I. & Sulistyoningrum, R. T. 2013. Menggali nilai-nilai keunggulan lokal kesenian reog Ponorogo guna mengembangkan materi keragaman suku bangsa dan budaya pada mata pelajaran ips kelas v sekolah dasar. *Premiere Educandum: Primary Education and learning journal of PGRI Madiun University* 3(1): 51–85.

Hadi, D.W. et al. 2019. *Statistik kebudayaan 2019*. Jakarta: Kementerian Kebudayaan dan Pendidikan Republik Indonesia (Ministry of Culture and Education of the Republic of Indonesia)

Nugraheni, W. 2018. Penanaman nilai-nilai moral melalui kesenian reog kendang terhadap pelajar di Kabupaten Tulungagung. *Imaji: Art and Education of Art Journal* 16(2): 162–171.
Rahayuningtyas, W. 2015. *Tari Topeng Malang*. Malang: State University of Malang (UM Press)
Rolitia, M. et al. 2016. *Nilai gotong royong untuk memperkuat solidaritas dalam kehidupan masyarakat kampung naga*. *Sosietas: Journal of Sociology Education* 6(1). 23–32.
Sangadji, E.M. & Sopiah. 2010. *Metodologi penelitian*. Yogyakarta: C.V Andi Offset
Santosa, E. 2017. Revitalisasi dan eksplorasi kearifan lokal *(local wisdom)* dalam konteks pembangunan karakter bangsa. *Journal of Diponegoro University Semarang.*
Sugiyono. 2018. *Metode Penelitian Kuantitatf Kualitatif dan R&D*. Bandung: Alfabeta
Suyatno, S. 2011. Revitalisasi kearifan lokal sebagai identitas bangsa di tengah perubahan nilai sosiokultural. *Journal of META SASTRA* 4(1): 82–89.
Widagdo, S. & Ermi D. K. 2014. Nilai pendidikan dalam upacara tradisi haul semangkin di Desa Mayong Lor Kecamatan Mayong Kabupaten Jepara. *LINGUA: Journal of Language and Literature* 10(1): 36–47.

Empowering Civil Society in the Industrial Revolution 4.0 – Sukowiyono et al (Eds)

Nyadran tradition as an effort to preserve Jenon's water resources and its socio-cultural values

W.D. Sulistyo* & N. Sadiyah
Universitas Negeri Malang, Malang, Indonesia

ABSTRACT: Javanese people appreciate local culture or wisdom. Especially if it is something that is disputed. One example of a community that still preserves ancestral culture is the people of Gunung Ronggo village, Tajinan subdistrict, Malang regency. The Community considers that the existence of Jenon water springs as sacred and worthy of preservation. A variety of traditions are done, one of them is Nyadran tradition which is routinely implemented in the Jenon water springs area. Not only as a preservation of the place, Nyadran tradition is also preserved because it contains positive social and cultural values. This is what caused the people of Gunung Ronggo village to preserve Nyadran's tradition as a form of preservation of the Jenon spring source.

Keywords: water, Sumber Jenon, sacred, belief, preservation, tradition, Nyadran

1 INTRODUCTION

Sumber Jenon is one of the local tourist attractions in Gunung Ronggo Village, T Crafts District, Malang Regency. This source is one of the water sites in the Greater Malang area (Sulistyo et al. 2020). Malang is an administrative area in East Java which has many historical sites (Sayono et al. 2019; Sulistyo et al. 2020) This place is actually a natural spring which then forms a pool of water. Unmitigated, this circular pool of water from Jenon Source measures about 650 m with a depth of up to 7 m. At the bottom of this pool, there is a large Jenu tree that collapsed decades ago. The local people believe that this tree is not an ordinary tree, but a sacred tree which then collapses and creates a mysterious spring. The spring is now called Sumber Jenon and is believed to have many benefits by the local community until now.

As a sacred place, of course Sumber Jenon is then preserved by the local community. Conservation itself is a process undertaken to manage natural resources, especially those that have certain benefits so that their authenticity is maintained at any time. This also shows that the role of the local community in the preservation of Sumber Jenon is very high. Now, Sumber Jenon spring is one of the tourist destinations for local and foreign visitors (Orbasli, 2002; Soeroso & Susilo, 2014).

One example of a form of preservation is the implementation of traditions. Tradition itself is a habit or custom carried out by the community from generation to generation from ancestors to today's society. Various traditions are also often held by the people of Gunung Ronggo Village as a form of respect for the sacredness of the Sumber Jenon area. One of the traditions routinely carried out at the Sumber Jenon spring complex is the Nyadran tradition. This tradition has always been celebrated jointly by all the villagers of Gunung Ronggo until now (Indonesia, 2002; Suwito et al. 2015).

The tradition of Nyadran is not without reason. There are several reasons why the people of Gunung Ronggo Village continue to maintain this tradition. One of them is the socio-cultural values contained in the Nyadran tradition. This tradition is an ancestral heritage that has primacy values

*Corresponding Author

 DOI 10.1201/9781003180128-24

(Sulistyo, 2016; Sulistyo & Pamungkas, 2020). As part of the millennial generation, cultivating characters such as socio-cultural values are very important. As time goes by, Indonesian people have begun to abandon the values of life which will gradually destroy this nation. Because the Nyadran tradition is a tradition that still contains efforts to instill character for the people of Gunung Ronggo Village, this tradition is still carried out regularly today (Yunus, 2013).

2 METHODS

The method used in the research is a qualitative research method. In the implementation process, the researcher collected data directly in the field with several activities, namely direct observation, interviews and documentation (Creswell & Creswell, 2017; Denzin & Lincoln, 2008), described in more detail below. This stage is carried out by the researcher in order to see firsthand the condition of the object under study. In this case the object in question is the Sumber Jenon tourist area which is located in Gunung Ronggo Village, T Crafts District, Malang Regency. This observation was carried out for one day on July 19, 2020. Researchers came directly to the observation area and noted important and interesting points in the Sumber Jenon area. The resource persons selected by the researcher in the interview process were Mr. Sholeh (28 years) as the manager of the Sumber Jenon tourist area and Mr. Aspuri (56 years) as the Head of Gunung Ronggo Village. Several things discussed with the manager of the tourist attractions and the Head of Gunung Ronggo Village were related to the conservation and management of Sumber Jenon

3 RESULTS AND DISCUSSION

3.1 *Site description*

The pool in the Sumber Jenon area is divided into two types, the main pool which contains the roots of the Jenu tree, and a small pool for small children to swim in. Based on research results from Asmaranto (Sulistyani & Irianto, 2018), the size of the main pond at Sumber Jenon is estimated to have an area of 649.1 m^2 with a depth of up to 7 m. In the main pool, it is also divided into two parts. One part that has a depth of up to 4 m is usually used for swimming for visitors, while the other part of the pool has a depth of up to 7 m and it is not allowed for visitors to swim in this area. The barrier for the two parts of the pool is a wooden bridge that visitors can walk through to get to the other side of the pond.

Apart from being a pool, the Sumber Jenon area is also surrounded by forests filled with very dense shady trees. Large trees such as Mahogany, Gondang, Banyan trees and so on still stand firmly around the pond. This makes the Jenon Source complex a beautiful complex that is not exposed to too much sunlight. Water sources are always considered by the local community as a sacred and sacred place, as well as for the people around the Sumber Jenon tourist spot, they consider this place sacred. There are still places for offerings around the pool area, used for visitors who want to pray or whoever has an interest.

According to Mr. Sholeh (28 years) as the manager of Sumber Jenon, before February 2019, the Sumber Jenon area was only an ordinary spring and not a tourist spot. This means that this place is not yet managed. This area is a village asset that is visited by the surrounding community. Several years ago, even if there was a large use, this place was only used as a gathering place for celebrations for local residents. Usually, every Raya day there is a thanksgiving in this area. Only in 2019, Sumber Air Jenon began to be managed as a tourist spot under the authority of Village-Owned Enterprises (BUMDes). As a place that has the potential to become an asset to the village's pride, of course many things are done by the community to conserve Sumber Jenon. One of the traditions that is still being carried out to preserve Sumber Jenon is the Nyadran tradition.

3.2 *Nyadran tradition*

Nyadran is one of the traditions that developed in Java since the past. During Hindu-Buddhist times, the Nyadran Tradition was known as the *Craddha* tradition, a form of ceremony to honor ancestors by bringing offerings and chanting praises during the implementation process. This *Craddha* tradition has at least been carried out since the Majapahit era, namely at the end of the 13th century. Meanwhile, during the Islamic period, the Nyadran tradition was also continued and continued to be carried out by including prayers from the Al-Quran. In fact, there are many sources that describe what the true function of Nyadran is. Sometimes, this tradition is also associated with the tradition of cleaning the tomb before Ramadan, the tradition of earth alms, or other traditions. Some people also believe that the Nyadran tradition is the result of a tradition of acculturation between Islam and Javanese culture. In essence, the Nyadran tradition in each place has a different function (Saddhono et al. 2019; Sih et al. 2018).

One group that still inherits this tradition is the people of Gunung Ronggo Village, T Crafts District, Malang Regency. In this area, there is one sacred place that is still sacred by the local community. That place is the source of Jenon. As a sacred place, of course the local people often hold events in the Sumber Jenon complex area. One of them is the Nyadran tradition. This tradition is carried out every *Selo* month *(Dzulhijjah)* every year. This means that this tradition is celebrated as a New Year's anniversary celebration. In addition, people often carry out the Nyadran tradition on every holiday.

Local people will come together to the Sumber Jenon spring area, bringing food which is commonly called *encek-encek.* While being led by traditional leaders and local religious leaders, the people of Gunung Ronggo Village will pray together hoping for the safety of the village and as an expression of gratitude. Then, people will exchange food before returning to their respective homes. In addition to the Nyandran tradition, there is also a local tradition called *Adat Kelapet.* This tradition is the tradition of protecting the seeds of the farmers from the river in this village. Implemented before the farming season, in hopes that the harvest will run smoothly. This tradition is similar to the *Wiwitan* tradition in Sentong Hamlet, Karangduren Village, Pakisaji District, Malang Regency. The local village community will carry out the *Wiwitan* tradition to pray that the harvest will run smoothly (Interview with Rosi, 21 years).

3.3 *Socio-cultural values in the Nyadran tradition*

As a tradition that is still routinely carried out until now, it can be said that the Nyadran tradition has benefits that are felt by the people of Gunung Ronggo. As with other traditions, if at present a tradition has no functional value for the people who carry it out, then slowly the tradition will be abandoned. Seeing that the Nyadran tradition is still ongoing in Gunung Ronggo village, it means that the local community thinks that this tradition has many benefits. Some things that can be seen from this tradition are its benefits in the social and cultural fields.

The Nyadran tradition in Gunung Ronggo village is also inseparable from the social and cultural values that make the local community still practice it today. Research was conducted to find out what social and cultural values are contained in these traditions. This research was conducted by making direct observations to the object of research, in this case the Village of Gunung Ronggo, and interviews with the Head of Gunung Ronggo Village and the Manager of the Sumber Jenon Spring Area. The results of this study are as follows:

First, the social values that exist in the Nyadran tradition. Social values can be interpreted as relationships between individuals. This relationship certainly has a positive impact between each other that makes each individual feel comfortable to continue the tradition in question. Social values also have many characteristics, including social interaction between communities, humanizing human nature, supporting harmony, and so on. In addition, social values also have a function as a unifying individual (Mustanir & Razak, 2017; Regan, 2000; Rohman & Hairudin, 2018). The social values that can be seen from the Nyadran tradition include the value of harmony and mutual cooperation. The value of harmony is not only an obligation for the Indonesian people, but also something that must be upheld in social life. In order to lead a harmonious life, each individual must also have a high awareness. This can be seen from the implementation of the Nyadran tradition in Gunung Ronggo Village. When this tradition took place, all village communities, whether young, old, male, female, Muslim, non-Muslim, all participated in the process of implementing Nyadran. Since ancient times, all the villagers of Gunung Ronggo have never been burdened by this tradition, they have been very active in participating in every implementation. This automatically shows harmony in the Gunung Ronggo community (Harahap, 2011).

Furthermore, there is the value of mutual cooperation. The value of mutual cooperation is also a very important value for social life. Mutual cooperation is very much needed for the welfare and advancement of society. By increasing the nature of mutual cooperation, at least this can also prevent conflict within the community. The value of mutual cooperation can also be shown through activities that are often carried out together, such as taking care of funerals, weddings, almsgiving the earth, celebrating holidays, and so on. In terms of this Nyadran tradition, the people of Gunung Ronggo show cooperative behavior in terms of big day celebrations. Together they coordinate how to carry out the Nyadran tradition with all kinds of preparations. Seeing that this tradition has been carried out for a long time and are still ongoing today proves that the people of Gunung Ronggo have practiced mutual cooperation from the past until now (Irfan, 2017; Korniawan, 2017; Slikkerveer, 2019).

Second, cultural values are also inseparable from the Nyadran tradition. This is because the Nyadran tradition is a tradition carried out by Javanese people since ancient times. Javanese people are known to highly uphold Javanese culture. Whatever culture was carried out by the ancestors, will still be preserved until now if it is felt that it provides benefits for them. Culture, tradition and Java cannot be separated. Indeed, the Javanese people cannot be separated from such traditions or beliefs. It can be said that the core of Javanese culture is belief itself. Therefore, the Javanese people tend to highly uphold existing traditions or myths (Mu'adi & Sofwani, 2018; Nawafil & Tiara, 2016).

Just like the Nyadran tradition for the people of Gunung Ronggo. This tradition is also upheld by all people as a form of respect for ancestral culture. Of course, the application of the Nyadran tradition has changed its function from what was previously a form of respect for Hindu Buddhists, to now a form of respect and great ceremony for the people of Gunung Ronggo who are predominantly Muslim. This proves that today, the Javanese people still value their ancestral culture, which can be proven by their availability to carry out various kinds of existing cultures.

4 CONCLUSIONS

The people of Gunung Ronggo Village are one of the groups of people who still hold sacred places. In this case the Sumber Jenon spring. Because this area is an area that is considered sacred, the villagers of Gunung Ronggo often hold big celebrations or traditions in the Sumber Jenon spring area. One tradition that is still very often carried out is the Nyadran tradition. Tradition is a form of tradition to honor ancestors by bringing food together and praying together in the Sumber Jenon area. This celebration is usually done in 1 Suro or other big events. Not only to preserve the holy spring of Sumber Jenon, the Nyadran tradition also has many social and cultural values. These values are always emphasized by the people of Gunung Ronggo Village to maintain their traditions. The examples of the values in question are the values of mutual cooperation, harmony and culture.

REFERENCES

Creswell, J.W. & Creswell, J.D. 2017. *Research design: Qualitative, quantitative, and mixed methods approach.* Sage publications.

Denzin, N.K. & Lincoln, Y.S. 2008. *Introduction: The discipline and practice of qualitative research.*

Harahap, S. 2011. *Teologi kerukunan*. Kencana.

Indonesia, T.R.K.B.B. 2002. *Kamus besar bahasa Indonesia*. Jakarta: Balai Pustaka.
Irfan, M. 2017. Metamorfosis gotong royong dalam pandangan konstruksi sosial. *Prosiding Penelitian dan Pengabdian Kepada Masyarakat* 4(1): 1–10.
Korniawan, R. 2017. Exploring the cultural values of "saemaul undong" and "gotong royong" as the development of an anti-corruption culture: socio-economic studies in East Asia and Southeast Asia. *Asia Pacific Fraud Journal* 3(2): 319–327.
Mu'adi, S. & Sofwani, A. 2018. Acculturation of Islam and Javanese culture in public servant ethics. *El Harakah* 20(2): 233.
Mustanir, A. & Razak, M.R.R. 2017. Nilai sosial budaya pada partisipasi masyarakat etnik Towani Tolotang dalam musyawarah rencana pembangunan. *Prosiding Konferensi Nasional Ke-6 Asosiasi Program Pascasarjana Perguruan Tinggi Muhammadiyah Aisyiyah (APPPTMA)*: 1–7.
Nawafil, M.I. & Tiara, N. 2016. Islam And Cultures: Java-Acculturated Islam in Islamic perspective. *International Conference on Social Politics*.
Orbasli, A. 2002. *Tourists in historic towns: Urban conservation and heritage management*. Taylor & Francis.
Regan, P.M. 2000. *Legislating privacy: Technology, social values, and public policy*. Univ of North Carolina Press.
Rohman, M. & Hairudin, H. 2018. Konsep tujuan pendidikan islam perspektif nilai-nilai sosial-kultural. *Al-Tadzkiyyah: Jurnal Pendidikan Islam* 9(1): 21–35.
Saddhono, K. et al. 2019. Local wisdom on Nyadran tradition in Sragen Regency. *INCOLWIS 2019: Proceedings of the 2nd International Conference on Local Wisdom, August 29-30, 2019.* Padang: EAI.
Sayono, J. et al. 2019. *Situs sejarah Malang Raya masa Islam dan kolonial*. Malang: Jurusan Sejarah FIS UM.
Sih, Y.N. et al. 2018. Religious value in nyadran ceremony in Ngepringan village, Sragen. *El Harakah* 20(2): 155.
Slikkerveer, L.J. 2019. Gotong royong: An indigenous institution of communality and mutual assistance in Indonesia. In *Integrated Community-Managed Development*. Springer.
Soeroso, A., & Susilo, Y.S. 2014. Traditional Indonesian gastronomy as a cultural tourism attraction. *Editorial Board*.
Sulistyani, K.F. & Irianto, D.B. 2018. Studi pemanfaatan air sumber Jenon untuk pemenuhan kebutuhan air irigasi dan domestik di kecamatan tajinan kabupaten malang. *Reka Buana: Jurnal Ilmiah Teknik Sipil Dan Teknik Kimia* 3(2): 137–142.
Sulistyo, W.D. et al. 2020. *Jesiamar: Jelajah situs Air Malang Raya*. Malang: Fakultas Ilmu Sosial.
Sulistyo, W.D. et al. 2020. *Jejak sejarah Malang Raya*. Malang: CV. Bintang Sejahtera.
Sulistyo, W.D. 2016. *Pengembangan model pembelajaran sejarah berbasis nilai-nilai serat wicara keras untuk meningkatkan kemampuan berpikirkritis siswa sma insan cendekia al mujtaba sukoharjo*. (Unpublished Thesis). Surakarta: Universitas Sebelas Maret.
Sulistyo, W.D. & Pamungkas, O.Y. 2020. Pemanfaatan situs sejarah peradaban Islam di Kota Malang sebagai aktivitas belajar untuk menanamkan nilai karakter. *AGASTYA: Jurnal Sejarah dan Pembelajaranya* 10(1): 1–15.
Suwito, S. et al. 2015. Tradisi dan ritual kematian wong Islam Jawa. *IBDA: Jurnal Kajian Islam dan Budaya* 13(2): 6–25.
Yunus, R. 2013. Transformasi nilai-nilai budaya lokal sebagai upaya pembangunan karakter bangsa. *Jurnal Penelitian Pendidikan* 13(1): 67–79.
Yusof, A. 2016. Relasi Islam dan budaya lokal: Studi tentang tradisi Nyadran di Desa Sumogawe Kecamatan Getasan Kabupaten Semarang. *IAIN Tulungagung Research Collections 4*(1): 67299.

Empowering Civil Society in the Industrial Revolution 4.0 – Sukowiyono et al (Eds)

Citizenship literacy for primary schools: An effort for Indonesia's future challenges

M. Wahono*
Universitas Negeri Semarang, Semarang, Indonesia

Dianasari
Universitas Muhammadiyah Cirebon, Cirebon, Indonesia

Y. Hidayah
Universitas Ahmad Dahlan, Yogyakarta, Indonesia

ABSTRACT: This study aimed to initiate citizenship literacy for elementary schools to discover solutions for Indonesia's future challenges. The method used qualitative methods by using several literature studies as a design. The results of the study showed that citizenship literacy for elementary school was done by using sites of citizenship that were customized to meet the needs of learning at the elementary school level. The understanding of citizenship literacy for elementary school was in line with the goal of Civic Education in the context of Indonesia. Besides, cooperation and efforts of various groups including family, government, community, and school were necessary to achieve the success of citizenship literacy for the elementary school in the future. This study recommends further research on some efforts in utilizing local wealth as an opportunity to release the golden generation in the future.

Keywords: citizenship literacy, children, elementary school, the future of Indonesia

1 INTRODUCTION

The optimization of literacy movement at an early age becomes an important thing to improve the output quality of education. However, the definition of literacy was not just limited to the ability of reading and writing, but also digital literacy beneficial for welfare (Fajriyah & Fauziyah, 2018; Rogers, 2011). The emphasis of the literacy movement at an early age was a good time to learn because there existed an opportunity to teach children about concepts (Hidayah et al. 2019).

Concerned the meaning of "literacy," Civic Education as a subject that has been taught since primary school to university as in accordance with the mandate of Article 37 paragraph (1) and (2) of Law Number 20 Year 2003 on National Education System, seeks to popularize citizenship literacy to form citizen in accordance with Pancasila and the Constitution of 1945. The introduction of the citizenship literacy concept in primary school is an attempt to provide a basis for students about the state: Indonesia.

The student learning environment was real-world participatory planning (Oonk et al. 2019). In other words, through citizenship literacy in elementary school, the habituation of thinking about Indonesia will be beneficial in higher education, even when they are working and raising a family. Regardless of age, a worker required the right skills and knowledge to become productive and to assist the organization in achieving strategic objectives (Davies et al. 2017).

*Corresponding Author

DOI 10.1201/9781003180128-25

At present, education attempts to address global challenges referred to as globalization. Globalization was a spacious social area, hegemonic, and intertwined with social groups on a world scale (de Sousa Santos, 2006). Presently, education is expected to produce individuals who are not only competent academically but to have critical and logical thinking. Therefore, it was beyond the current power configuration problems and tended to shift power (Rizvi, 2007).

Literacy education in primary school can be started by reading habituation. There were many different approaches to teach literacy at the early ages (Hill, 1995). The levels of literacy skills are along with children's ability to receive, process, and respond to any information. A study that examined the effects of early literacy teaching in kindergarten students showed that children who received additional instruction performed significantly better in terms of rhyme, alliteration, knowledge of letters, the relationship of letter and sound, spelling, and combine tasks than children who receive instructions one by one through a tutoring program (Bingham et al. 2010).

Based on a report from the National Early Literacy Panel in 2008, there were at least extensive meta-analyses of 300 studies that showed early literacy measures that correlated with later literacy achievement (Shanahan & Lonigan, 2010). Thus, the early stages of literacy are fundamental in fostering the critical thinking and logical nature of children.

The previous research that discusses the level of citizenship literacy in elementary school is: a study in Korea for two years on ten fifth and sixth grade students (11–12 years old) that proposed to incorporate intercultural critical citizenship, illustrated those critical skills of citizenship can be developed with the pedagogy of teachers. They could help students to reflect the knowledge and cultural practices that seem natural to them (Huh & Suh, 2017).

A study on promoting popular media as a pedagogical tool institutionalized within the framework of critical media literacy that refers to values and codes of multiculturalism, found a promoted model of citizenship in students' textbooks in 2003 composed by the Ministry of Education Ontario in Canada. On the contrary, the promotion failed. The curriculum of Ontario media of literacy failed to interrogate social roots of conflict and discrimination (Ferguson, 2011).

An analysis that aimed to bring the theory of intercultural civic education becomes a concern among academics, whereas civic education explained the reasons and concepts involved in intercultural Citizenship Education (Porto et al. 2017). Academically, Walsh et al. (2014) gave evidence to overcome factors outside the school that affect students' learning. By conducting research on 7.948 students from 1999 to 2009, has resulted in students at *City Connects* presenting a high report score, and get a high score in Arts, Math, and English after giving a supportive intervention for them.

Furthermore, a study that described some of the core aspects of the experience of Omar Dengo Foundation at Costa Rica in the development of Citizenship Education program based on the conception of children as a citizen have produced that the educational program becomes a modeled program designed for elementary and it presented lessons learned from the process of improving initiatives (Fonseca & Bujanda, 2010).

Based on previous studies, we know that instilling any concept in Citizenship Education has been implemented through many challenges from time to time. The argument makes researchers interested in conducting a study on Citizenship Literacy for Elementary School to prove that students at the elementary school level have a strategic position for citizenship concept instilling. The subject was 8 years old students and 6 years old students that became a mental age predictor of success (Lutzer, 1986).

The previous research has demonstrated the positive impact of doing research on teacher's pedagogy. However, there are a few studies on the challenges and benefits of doing research in Indonesia, the Philippines, and ASEAN regions (Ulla, 2018). Thus, this study formulates a question on "How can citizenship literacy in elementary school deal with facing challenges of Indonesia in the future?"

The position and importance of this study are: it is theoretically useful to add insight into citizenship at the primary school level and it has practically become the basis for similar further studies. This study contributed to scientific development and the resolution of related problems regarding civic education in primary schools.

2 METHODS

This study used qualitative descriptive in terms of literature study as design. Data was collected from various sources such as books, journal articles, and other relevant sources. Knowledge embedded in the value system of individuals was the fundamental basis for explaining how innovation occurs (Bressan, 2004). The researchers compared some data to determine citizenship literacy for elementary schools to answer Indonesia's challenges in the future. Researchers compared several data to determine the nationality of primary schools to respond to Indonesia's challenges in the future. In detail, the stages of implementing library research were: (1) keywords search on Citizenship Literacy for Elementary Schools, (2) search finding the latest books and scientific articles on Citizenship Literacy for Elementary Schools, (3) systematic reading analysis, especially full-text sources on Citizenship Literacy for Elementary Schools, (4) grouping Citizenship Literacy for Elementary Schools to find the core of the problem, (5) determining conclusions and suggestions.

3 RESULTS AND DISCUSSION

3.1 *Site of citizenship literacy for elementary school*

The idea of citizenship sites as citizenship literacy sites for elementary school is an effort to exploit these sites to be citizenship literacy for primary school (elementary school). The concept of citizenship site was an initiative at the local level that allows for democratic citizenship practices that could encourage citizen participation in the context of the nation's life (Muwafiq, 2012).

At the elementary school level in Indonesia, citizenship literacy can be started by introducing the state, state's motto, state's principle. Many early literacy interpretations were too narrow and ignore the important role of background knowledge and conceptual development (Neuman, 2010). The introduction in the level of state's structures at the Primary School level in Indonesia is enabled to prepare students' social and emotional development, cognition aspect, language, and citizenship literacy on a number of important aspects that are compulsory to be possessed by students as Indonesian.

The urgency of children's preparation as Indonesian is in line with the objectives of National Education stated in Law Number 20 Year 2003 which ends with the assertion that each democratic citizen need to "...become a democratic and responsible citizen." The development of children's citizenship literacy concerns existing patterns of education.

The site of citizenship as citizenship literacy for elementary school expects the existence of practical improvement of citizenship in maintaining the existence of citizens in democratic countries. In America, there was an idea that the current American schools' crisis is a symptom of a wider crisis in the meaning and practice of democracy (Giroux, 1992). In Indonesia, the site of citizenship as citizenship literacy for elementary school refers to the introduction of national identity maintenance. A study that examined how character education and multicultural values education can contribute to Indonesia in 2045 (100 years old of Indonesia), produced that the form of education should not be limited to conceptualization aspect, but to understanding aspect of personality traits and considering globalization aspect (Malihah, 2015).

For Indonesian, national identity is a manifestation of national identity in the future. The elementary school as formal education in Indonesia has an opportunity to take a part in nation intelligence development as stated in "Unity in Diversity" or *Bhinneka Tunggal Ika* at the elementary school level. Citizenship education should be reformed to reflect home cultures and languages of students from various groups, and this action could help individuals to achieve structural equality (Banks, 2008).

The pattern of recognition on citizenship sites as citizenship literacy for Elementary School is to emphasize the fact of Indonesia as a great nation with a wealth of language differences, religion, and culture as Indonesian's national identity. Citizenship literacy at the elementary school level is expected to have a positive impact on the introduction of the statehood (Indonesia) concept on students.

3.2 *Citizenship literacy for elementary school and future challenges of Indonesia*

The concept of citizenship literacy for elementary school also becomes an attempt to prepare children to have the ability to learn citizenship. The movement of critical thinking was expected to benefit from conceptualizing the lesson material in a development framework (Kuhn, 1999). Concerning the challenges of Indonesia in the future, the 21st-century curriculum is to consider the use of information technology and the emphasis on graduates that are ready to work. Besides, discipline forced the 21st-century curriculum to prepare graduates to learn continuously (Gosper & Ifenthaler, 2014). There were at least seven challenges for Indonesia in the future that could be formulated, i.e. a demographic bonus, stunting, qualified early childhood education, dropout rates, unemployment, disruption of technology, and regional autonomy (Harususilo, 2019). In the 21st century, the pattern of education is expected to increase. There was a suggestion that cultural awareness and background knowledge of students could eliminate competence hamper students' learning (Wang, 2018).

To respond to the challenge, the problem of attitude and way of communication between teacher and student has become a concern. A study that examined the role of social support and adjustment in predicting academic stress among first-year students at Syiah Kuala University revealed that the two aspects of social support and customization significantly predicted academic pressure (Hamzah & Marhamah, 2015). The present education is required to be able to adapt with times, especially in the era of the industrial revolution.

Citizenship literacy for elementary school and Indonesia's future challenge is an effort to prepare superior human resources in the future based on Pancasila and the Constitution of 1945. The study on education was distracted by a skeptic that doubted its quality and relevance (Ball & Forzani, 2007). For elementary schools, understanding aspects of Indonesia is a solution to connect information with other useful information of Indonesia.

To reconsider the challenge of Indonesia in the near future, citizenship literacy for elementary school occurs as a perspective in finding out solutions on civic knowledge degradation in Indonesia. The attitude of lateral preference on children was not dimensional (Gudmundsson, 1993). It is crucial to understand the position of education in Indonesia to respond to the symptoms of civic knowledge degradation. In the last decade, the important struggles on the production of knowledge had occurred in history, social sciences, and education (Popkewitz, 1997).

The composition of an educational model that can export competence of learners is used to formulate competence in the 21st century. The broad generalization in all aspects of human psychology concerns the major differences of individuals. On the other hand, the strength, sturdy, and well-established human capabilities were not utilized well in school (Suppes et al. 1998). The presence of learning media in the industrial revolution is appropriately utilized to support the learning process.

Citizenship literacy for elementary school and Indonesia's future challenge is formulated to infiltrate into various fields of learners. The distribution of education output was focused on soft skills (Hariyanto et al. 2019). Any change runs continuously. Thereby, citizenship literacy for the elementary school in Indonesia is in an offer to address challenges of the future.

4 CONCLUSIONS

Citizenship literacy for elementary school was to confirm the purpose of Civic Education in the context of Indonesia. Cooperation of family, government, community, and school was expected for the success of citizenship literacy in elementary school. Citizenship literacy for elementary school was a basis for students to recognize civic activity. Therefore, to support the realization of a smart and good citizen, citizenship literacy needs to be popularized by academics of Civic Education in elementary school. Future studies are necessary to analyze local wealth as an opportunity to release the golden generation in the future.

REFERENCES

Ball, D.L. & Forzani, F.M. 2007. Wallace Foundation Distinguished Lecture-What makes education research "educational"? *Educational Researcher* 36(9): 529–540.

Banks, J.A. 2008. Diversity, group identity, and Citizenship Education in a global age. *Educational Researcher* 37(3): 129–139.

Bingham, G. E. et al. 2010. Systematic and engaging early literacy: Examining the effects of paraeducator implemented early literacy instruction. *Communication Disorders Quarterly* 32(1): 38–49.

Bressan, B. 2004. *A study of the research and development benefits to society resulting from an International Research Centre.* Helsinki: Faculty of Science of the University of Helsinki.

Davies, E. et al. 2017. Learning and training for older workers. In H. Karen et al. (eds), *Managing the ageing workforce in the East and the West*: 185–206. United Kingdom: Emerald Publishing Limited.

de Sousa Santos, B. 2006. Globalizations. *Theory, Culture & Society* 23(2–3), 393–399.

Fajriyah, L. & Fauziyah, P. 2018. The role of ICT to develop early literacy of children in era digital. In R. Windiarti (ed.), *Semarang early childhood research and education talks (SECRET 2018)*: 112–116; *Proc. intern. conf., Semarang, 8–9 August 2018.* Paris: Atlantis Press.

Ferguson, S. 2011. Classroom contradictions: Popular media in Ontario schools' literacy and citizenship education policies. *Education, Citizenship and Social Justice* 6(2): 137–151.

Fonseca, C. & Bujanda, M.E. 2010. Promoting children's capacities for active and deliberative citizenship with digital technologies: The CADE Project in Costa Rica. *The ANNALS of the American Academy of Political and Social Science* 633(1): 243–262.

Giroux, H. A. 1992. Educational leadership and the crisis of democratic government. *Educational Researcher* 21(4): 4–11.

Gosper, M. & Ifenthaler, D. 2014. *Curriculum models for the 21st Century: Using learning technologies in higher education.* Netherlands: Springer.

Gudmundsson, E. 1993. Lateral Preference of Preschool and Primary School Children. *Perceptual and Motor Skills* 77(3): 819–828.

Hamzah, H. & Marhamah, F. 2015. Social support, adjustment and academic stress among first year students in Syiah Kuala University. In S.I.A. Dwiningrum (eds), *2nd international conference on current issues in education*: 447–450; *Proc. intern. conf., Yogyakarta, 25–26 August 2015.* Yogyakarta: Yogyakarta State University.

Hariyanto, H. et al. 2019. The analysis of student's verbal communication skills by gender in the middle school in South Tangerang. In A. Suparmi & D.A. Nugraha (eds), *International conference on science and applied science (ICSAS)*: 020064-1–020064-6; *AIP conf. proc., Surakarta, 20 July 2019.* Surakarta: AIP Publishing.

Harususilo, Y.E. 2019. *7 tantangan pendidikan indonesia, dari stunting sampai pengangguran.* Retrieved from kompas.com.

Hidayah, Y. et al. 2019. Transformasi kearifan lokal jawa dalam pendidikan karakter sekolah dasar. *AULADUNA: Jurnal Pendidikan Dasar Islam* 6(1): 50–61.

Hill, S. 1995. Early literacy and diversity. *Australasian Journal of Early Childhood* 20(1): 23–27.

Huh, S. & Suh, Y.M. 2017. Preparing elementary readers to be critical intercultural citizens through literacy education. *Language Teaching Research* 22(5): 532–551.

Kuhn, D. 1999. A developmental model of critical thinking. *Educational Researcher* 28(2): 16–46.

Lutzer, V.D.F. 1986. Perceptual learning by educable mentally retarded, average, and gifted children of primary school age. *Perceptual and Motor Skills* 62(3): 959–966.

Malihah, E. 2015. An ideal Indonesian in an increasingly competitive world: Personal character and values required to realize a projected 2045 'Golden Indonesia'. *Citizenship, Social and Economic Education* 14(2): 148–156.

Muwafiq, A. 2012. Pesantren sebagai situs kewarganegaraan dan pembentukan budaya kewargaan di Madura. *Jurnal Civics* 14(2): 183–195.

Neuman, S.B. 2010. Lessons from my mother: Reflections on the National Early Literacy Panel Report. *Educational Researcher* 39(4): 301–304.

Oonk, C. et al. 2019. Educating boundary crossing planners: Evidence for student learning in the multistakeholder regional learning environment. *Journal of Planning Education and Research* 39(3): 360–373.
Popkewitz, T. S. 1997. A changing terrain of knowledge and power: A social epistemology of educational research. *Educational Researcher* 26(9): 18–29.
Porto, M. et al. 2017. Intercultural citizenship in the (foreign) language classroom. *Language Teaching Research* 22(5): 484–498.
Rizvi, F. 2007. Postcolonialism and globalization in education. *Cultural Studies ↔ Critical Methodologies* 7(3): 256–263.
Rogers, T. July 2011. *Literacy, leadership and marginalized groups*. In the 7th Pan African Reading for all Conference, Gaborone: University of Botswana.
Shanahan, T. & Lonigan, C. J. 2010. The national early literacy panel: A summary of the process and the report. *Educational Researcher* 39(4): 279–285.
Suppes, P. et al. 1998. "The vision thing": Educational research and AERA in the 21st century – Part 5: A vision for educational research and AERA in the 21st Century. *Educational Researcher* 27(9): 33–35.
Ulla, M.B. 2018. Benefits and challenges of doing research: Experiences from Philippine public-school teachers. *Issues in Educational Research* 28(3): 797–810.
Walsh, M. E. et al. 2014. A new model for student support in high-poverty urban elementary schools: Effects on elementary and middle school academic outcomes. *American Educational Research Journal* 51(4): 704–737.
Wang, G. 2018. On the strategies of enhancing students' cultural awareness in College English Teaching. *English Language Teaching Canadian Center of Science and Education* 11(12): 116–120.

Empowering Civil Society in the Industrial Revolution 4.0 – Sukowiyono et al (Eds)

Pancasila and multicultural values to strengthen citizens ethics in encountering global challenges

S. Dharma*, Hodriani, R. Nababan & S. Hadiningrum
University State of Medan, Medan, Indonesia

R.J. Moertiono
Muhammadiyah University of North Sumatra, Medan, Indonesia

ABSTRACT: This study examined the ethics of citizens facing various global challenges internalizing the value of Pancasila and respecting others as a multicultural nation-state, Indonesia. Applying a qualitative approach, this study collected data from experts in the field of civics, lecturers of Pancasila and Civic Education in several universities in Indonesia. The results showed that a joint commitment was needed in solving various global challenges regardless of ethnicity, race, or religion. In order to strengthen the citizen's commitment as a nation, it was necessary to internalize the values of Pancasila and the fact that Indonesia was a multicultural nation. The values that need to be internalized have three dimensions: personal dimension, interactive dimension, public dimension, and social dimension.

Keywords: Pancasila, multiculturalism, citizen ethics, global challenges

1 INTRODUCTION

The world is currently facing various global challenges that affect social life both at the local and national levels. Various global issues such as environment, humanity, health, economy, and politics that occur in one country will affect other countries. As with various traditional and non-traditional global issues which are currently experiencing strengthening and having an impact on world life (Winarno 2014). Currently, about one point one billion people are at high risk of being affected by violent conflicts that occur in the future. In fact, most of the violence occurs intra-state, between civilians, not inter-state, between countries (Bhargava 2019). The International Institute for Strategic Studies (IISS) states that due to the global conflict throughout 2015 the death toll has reached 167,000 (BBC 2019). Likewise, with the Covid-19 virus which initially developed in Wuhan China, now has an impact on almost all regions in the world including Indonesia. The question is what kind of ethics citizens should have to face various global challenges, what are the values contained in Pancasila that can be used as guidance in facing various challenges, both local, national and global? The multicultural values of the Indonesian people can be used as a reference to behaving and acting like citizens.

Pancasila is the foundation of the Republic of Indonesia. The values of Pancasila have existed long before the Indonesian nation founded the state. These values come from the customs, culture, and religious values of people who lived long in the archipelago region before it was called Indonesia (Kaelan 2013). Further, when Indonesia is founded, these values are inherent in everyday life as a way of life for the nation. Furthermore, the founders of the state explored and formulated these values to become the basis of the Indonesian state philosophy (Alqadri 2009). There are various

*Corresponding Author

DOI 10.1201/9781003180128-26

essential values contained in Pancasila, namely Divinity, Humanity, Unity, Population, and Justice. These values objectively serve as guidance of the Indonesian people in thinking, acting, and behaving. These values encourage social awareness in the form of moral obligations as a consequence in a pluralistic society. These moral consequences then give birth to attitudes, characters, and behaviors to exploit and increase the positive potential by reducing negative prejudices that hinder change for future progress (Cameron 2018). In facing various global challenges, values are needed as the basis of thinking and acting. These values then become the ethics of citizens to take part in facing and solving various global challenges.

To achieve the human welfare of the twenty-first century, a global ethic is needed, so that everyone can act based on their obligations as a human being regardless of nationality or identity. Everyone has a moral imperative with various skills, knowledge, and values that can be used to act in an increasingly global society (Convery & Kerr 2007). Because there are various global problems and challenges that must be resolved collectively based on the values that are believed to guide behavior and actions. In the Indonesian context, these values come from Pancasila and Multicultural values. This paper examines the urgency of Pancasila and the multicultural values of the Indonesian nation in strengthening the ethics of citizens facing global challenges. This paper will examine what universal values can be used as guidelines in attitudes and actions to face global problems and challenges.

2 METHODS

This study applied interviews as the main technique to collect data. The interviews obtained data about the thoughts and views of respondents on the topic of this research. The respondents involved are experts in the field of civics and lecturers of Pancasila Education and Citizenship Education. The research was conducted at Indonesian Education University, Medan State University, and Yogyakarta State University. In addition to the interview, this study included group discussions (FGD) and documentation studies. Qualitative data analysis of Milles and Huberman (2007) was applied to stream of data reduction, data presentation, and drawing conclusions/verification.

3 RESULTS AND DISCUSSION

3.1 *Joint commitment in resolving various global challenges*

The citizens of the world are members of the human community that have responsibilities beyond the boundaries for their nation and state. According to Noddings (2005) and Chong (2015), global citizens who care about the economy, social justice, the earth, social diversity, culture, and peace are those who involve identity, loyalty, or commitment beyond the borders of their nation.

This opinion is in line with what was stated by respondents through interviews which explained that:

> "*Every citizen has a responsibility in responding to various global challenges. To strengthen this responsibility, a number of skills are needed. Those skills are social responsibility, which is understanding inequality, global justice, empathy, and caring, and personal responsibility, and the interrelation of global problems; global competence, related to why it is important that global competencies are given to students. Global competencies such as self-awareness, intercultural communication, and global knowledge; involvement as global citizens, namely the values that will arise as a result of the involvement of students in community organizations. For example, at the international level, political aspirations, activism as a global citizen.*"

What the respondents said above was in line with what was stated by other respondents, "Because we are in a global era, we must always be able to think globally even though we have to act locally. For this reason, the role of global citizens must be strengthened in everyone, including students". Adding to that, another respondent said, "To become a proactive citizen, the responsibility as a global citizen must be cultivated to become citizens who behave fairly, fairly, peacefully, tolerant, inclusive, safe and sustainable".

From the opinions of those respondents, it can be understood that "Everyone has a responsibility to solve various global problems and challenges. Like it or not, global trends will affect the lives of citizens, so that everyone has an obligation to be involved in it". As citizens, everyone is called to understand various issues such as agriculture, human rights, ecology, and so on which ultimately unites people from various countries (Nussbaum 2002). To achieve this goal, three capacities are needed by citizens, namely ability to perform critical self-examination, ability to think as citizens of the world, and narrative imagination, namely the ability to think what it feels like in the shoes of people who are different from oneself.

In this regard, it can be understood that global problems and challenges are the responsibility of everyone regardless of ethnicity, ethnicity, religion, or even nation. Because humans are actually global citizens who must learn how to recognize the humanity of all humans (Enslin 2011). The attitude of trans-national solidarity will encourage faster resolution of various global problems and challenges. Moreover, this attitude of solidarity is formed and comes from the cultural values of each nation.

3.2 *Citizen ethics sourced from Pancasila and the multicultural values of the Indonesian nation*

For Indonesia, the presence of Pancasila is the nation's greatest gift. The values of Pancasila are the deepest dimension which is the source of ethics and morals of the Indonesian nation which then develops in the form of attitudes and behaviors or real actions in the lives of citizens (Yuliarto et al. 2020). As a source of ethics and national morals, Pancasila is used as a guide for thinking and acting in solving various global problems and challenges. This is in line with what was stated by respondents who stated that "*The values of Pancasila must be used as the basis for strengthening responsibility and global awareness in citizens*". This opinion is supported by other respondents who stated that "*Indonesia is a very plural country bound by the Pancasila Ideology. For that, how do we now make Pancasila a guide in thinking and acting both at the local, national, and global levels*?"

For example, how the fourth principle of Pancasila, namely "*Social Justice for All Indonesian People*", can be interpreted that social justice obliges countries to think about the welfare of humanity (Kaelan 2013). Social justice regulates the relationship between countries and the world community or vice versa which is based on efforts to uphold the basic rights of mankind (Kaelan 2013).

Apart from Pancasila, Indonesia has virtuous values that come from various customs, cultures, and religions in Indonesia. Currently, there are more than 1300 ethnic groups in Indonesia and 6 religions recognized by the state. Every ethnicity and religion have noble values that can serve as guidelines for behavior. These values are universal which govern the relationships of everyone in social life. For example, justice where every tribe, ethnicity or religion respect that value. Furthermore, compassion, namely there is no nation that does not support the value of compassion. As well as caring, respecting, etc. which are also basic values in every ethnicity, ethnicity or religion.

From this explanation, the ethics of citizens in facing various global problems and challenges are universal values that serve as guidelines in the thinking and attitude of the Indonesian nation which are described below.

Figure 1 shows the 4 dimensions of values that are the source of extracting universal multicultural values. Universal concerns the development of a shared (collective) identity and embracing the same values of all mankind (respect for human life, environmental preservation, and others). The universal values referred to are values that can be understood and accepted by all ethnic groups, races, cultures, and religions in Indonesia. These four values are at the same time the basis of humans to make conscious decisions based on human instincts and emotions (Cameron 2018).

Values that have a personal dimension are values that come from within (conscience) so that they are able to distinguish between right and wrong. Examples of characters that have a personal

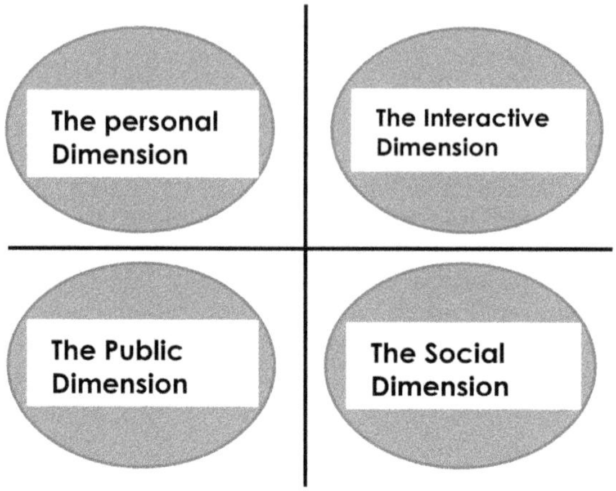

Figure 1. Four dimensions that become the source of extracting multicultural values.

dimension are religious, honest, loyal, empathetic, conscious, etc. The interactive dimension is the values that a person uses to establish relationships (interactions) with other people. Examples of these values are responsibility, discipline, desire to do your best, friendship, etc. Values with a social dimension, namely the values that are used by everyone in carrying out their life in the midst of society. Examples of this value are mutual cooperation, politeness, respect and respect for differences (tolerance), harmony, social care, etc. Finally, values have a public dimension, namely universal values that are recognized, practiced, and fought for by all people in the world. Examples of these values are equality, justice, sustainability, caring for the environment, loving peace, democracy, etc.

These values become a reference for citizens in thinking and acting when facing various global problems and challenges. For example, the world is currently facing armed conflict that has resulted in civilians becoming victims. Armed conflict is very contrary to the values that exist in Pancasila as well as the multicultural values of the Indonesian as a nation that want peace, compassion, mutual respect for differences, and so on. Likewise with the environmental crisis due to global warming. Everyone can reflect on how the teachings originating from Pancasila and the customs, culture, and religion in Indonesia teach us to love nature without destroying it. When these values are implemented, everyone will realize that how to treat others will have implications for the meaning of justice for all mankind (Enslin 2020).

4 CONCLUSIONS

The world is currently facing various global challenges that affect people's lives both at the local and national levels. For that, we need collective efforts of citizens to solve it. One effort that can be done is to strengthen the Pancasila and multicultural values of the Indonesian nation.

The values of Pancasila and the multiculturalism of the Indonesian people are universal values that serve as guidelines in thinking and acting by citizens in responding to various global problems and challenges. These universal values originate from the teachings of customs, culture, and religion in Indonesia. These values are then used as citizen ethics in behaving and acting. The value in question is (1) values that have a personal dimension (care and awareness, etc.); (2) value with interactive dimension (responsibility, etc.); (3) values that have a public dimension (fairness, equality, etc.); and (4) values with a social dimension (mutual cooperation, etc.).

So, a joint commitment is needed in solving various global challenges regardless of ethnicity, ethnicity, race, or religion to strengthen this commitment, it is necessary to internalize the values that are derived from Pancasila and the Multicultural values of the Indonesian nation which include four dimensions, namely values with a personal dimension; interactive dimension value; values that have a public dimension; as well as values with a social dimension

REFERENCES

Alqadri, S.I. 2009. Pluralisme dan multikulturalisme dalam pemeliharaan kerukunan umat beragama. In *Memelihara Kerukunan Melalui Pendidikan Multikultural*: 13–34. Jakarta: Kedeputian Bidang Koordinasi Pendidikan, Agama dan Aparatur Negara.

BBC. 2019. *Korban akibat konflik global mencapai 167.000 jiwa*. Retrieved from https://www.bbc.com/indonesia/dunia/2016/05/160505_dunia_perang

Bhargava, V. 2006. Introduction for global issue. In B. Vinay (Ed.), *Global issue for global citizens an introduction to key*: 1–30. Washington: The World Bank.

Cameron, J.D. 2018. Communicating cosmopolitanism and motivating global citizenship. *Political Studies*, 66(3): 718–734.

Chong, E.K. 2015. Global citizenship education and Hong Kong's secondary school curriculum guidelines. *Asian Education and Development Studies* 4(2): 221–247

Convery, A. & Kerr, K. 2007. Acting locally to have global impact: Citizenship Education in theory and practice in England. *Citizenship, Social and Economic Education* 7(3): 189–200.

Enslin, P. 2011. Education for global citizenship: The cosmopolitan and the patriotic. *Citizenship, Social and Economic Education* 10(2): 91–10.

Kaelan, M.S. 2013. *Negara kebangsaan Pancasila: Kultural historis, filosofis, yuridis dan aktualisasinya.* Yogyakarta: Paradigma.

Miles & Huberman, A. 2007. *Analisis data kualitatif: Buku sumber tentang metode-metode baru.* (T. Rohendi, Trans). Jakarta: Universitas Indonesia Press.

Noddings, N. 2005. *Educating citizens for global awareness.* New York: Teacher College Press.

Nussbaum, M. 2002. Education for citizenship in an era of global connection. *Studies in Philosophy and Education* 21: 289–303.

Winarno, B. 2014. *Dinamika isu-isu global kontemporer*. Yogyakarta: Center of Academy Publishing Servicer (CAPS).

Yuliarto, A. et al. 2020. *Bidang studi empat konsensus kebangsaan*: *Sub bidang studi Pancasila.* Jakarta: Lembaga Ketahanan Nasional Republik Indonesia.

Empowering Civil Society in the Industrial Revolution 4.0 – Sukowiyono et al (Eds)

A conceptual learning model of global civic education in Indonesia

Sutrisno*
Citizenship Education, Universitas Muhammadiyah Ponorogo, Ponorogo, Indonesia

Sapriya, K. Komalasari & Rahmad
Universitas Pendidikan Indonesia, Bandung, Indonesia

ABSTRACT: This study aimed to analyze the conceptual learning model of global civic education in Indonesia. This research used mixed methods. Data collection techniques were conducted through questionnaires and interviews. The research used a stepwise qualitative-quantitative data analysis. The research was conducted on students and teachers of Pancasila and Civic Education in senior high school. The results showed that the application of the concept of global civic education in Indonesia could be actualized through the learning model of a global citizens project. A global citizens project learning model is a learning model that develops the dimensions of global citizenship of young people in critical thinking and problem solving, communication and collaboration as well as creativity and innovation in solving issues at the local, national and global levels which include the impacts of climate change, social culture, legal politic, economic, and education.

Keywords: global citizenship, learning model, Pancasila and civic education, young people

1 INTRODUCTION

Global problems related to local and global issues are shared problems, part of all nations throughout the world (Harshman & Augustine, 2013). Therefore, the concept of learning at this time must contribute in shaping the skills of 21st century citizens in an effort to prepare citizens with global perspectives who can contribute in solving problems of global and local issues.

The results of Karsten et al. (1998) also showed that there were seven global trends that needed to be addressed by each country. They were: (a) the economic gap between countries will be significantly expanded and sustainable, (b) information will increase significantly, (c) the increasing differences between those who have and lack technology, (d) conflicts between interests in developed and developing countries are increasing, (e) financing access to clean water had increased dramatically due to environmental damage and population growth, (f) illegal logging or dramatic deforestation affecting the diversity of quality of life, (g) population growth in developing countries resulting in populations, especially children, living below the poverty line.

In line with this, the results of Murdiono's research also showed that many global problems that occurr today were certainly caused by the role of young citizens in understanding the nature of low global citizenship, especially in the learning process of civic education in schools and even colleges (Murdiono, 2014). This is also supported by the increasingly widespread dissemination of technology and information among the public. Information and communication technology (ICT) was utilized as if there was no limit to accessing it, without knowing the area and time. However, ICT could also be used as a medium for exchanging information and ideas between citizens and government (Kim et al. 2011).

*Corresponding Author

 DOI 10.1201/9781003180128-27

Based on field findings and observations at the levels of senior high school, vocational high school, and *madrasah aliyah*, the global outlook of students was also limited to a basic understanding of the concept of problem-solving materials and global issues; it was still not a direct practice (Sutrisno, in prep.). The dimension of global citizenship students consisted of several aspects i.e., global responsibilities, global competence, and global participation which were still low both in the form of basic understanding and practice. Meanwhile, the concept of a learning model which was oriented towards improving learning skills and innovating students (critical thinking and problem solving, communication and collaboration, creative and innovative) was quite well applied. But it had not been fully developed in the form of social activities in an effort to resolve global and social issues. The concept of 21st century learning had also not been fully developed in the learning process in the classroom by educators.

Table 1. The questionnaire results of a conceptual learning model in the global citizens project for developing learning skills and innovating learners of 21st century students.

No	Indicator	Results		
1	Global citizenship dimension	Global responsibilities	Global competence	Global participation
	Research object source (90)	40%	35%	40%
	Research outside object source (90)	45%	40%	40%
2	Learning and innovating skills	Critical thinking	Communication collaboration	Innovative creative
	Research object source (90)	50%	50%	45%
	Research outside object source (90)	55%	50%	50%

These problems certainly must be resolved not only through legal products or government policies, but must be done early, especially in the education world. The world of education is the basis for applying the habituation of young people in building the character of national identity in encountering the developments in the 21st century (Sutrisno, in prep.). Thus, there needs to be a reconstruction of the learning model to contribute to the resolution of issues and problems at the local, national, and global levels.

Global civic education is an educational program discourse developed by UNESCO in an effort to participate in solving various problems, issues, and conflicts globally through the learning process both in the classroom and outside the classroom (UNESCO, 2015). The application of the global civic education model for each country has differences in terms of both the approach and implementation program. There are several mappings in the implementation of global civic education in various countries. However, the concept of global civic education will be effective when each country is given a separate policy regarding the implementation of the civic education model that will be used (VanderDussen Toukan, 2018).

Thus, we need a conceptual model of the global civic education conceptual model which is ideally developed and implemented in Indonesia, especially at high schools. The concept of this learning model must be oriented to the philosophical basic values of the Indonesian people, which are based on Pancasila as the nation foundation in solving various problems, both at the local, national, and global levels (Murdiono, 2014).

2 METHODS

This research used a mixed methods research approach. This research used an explanatory sequential method. The data sources consisted of students and teachers of Pancasila and Civic Education at senior high school/vocational high school/*madrasah aliyah* levels in Ponorogo regency and outside Ponorogo regency covering various regions in Indonesia. This study used cluster random sampling and took five months. In Ponorogo regency, determination of research locations was based on the level of superior schools. Meanwhile, the data from outside Ponorogo regency was used as a comparison.

The first stage of this research used a quantitative method i.e., a survey about the conceptual description of the learning model of the global citizen's project. The questionnaire was used to see the extent to which the learning components of global citizenship education which include global responsibility, global competence, global participation, and learning and innovation skills have been implemented in the current learning process.

The second stage used a qualitative method. At this stage, the interview with teachers was conducted to find out the conceptual description of the implementation of global civic education learning in Pancasila and Civic Education learning in various schools. The results of this mix of methods aimed to create a conceptual model of global civic education. This conceptual final result will be developed into a prototype learning model in subsequent studies.

3 RESULTS AND DISCUSSION

3.1 *Concepts and theories of the global citizens project learning model as the implementation of global civic education in Indonesia*

Global civic education is a form of civic education that functions as controlling part of the education system in the 21st century era which is full of social problems. The education system has to be prepared early on, especially in developing the character of national identity, in facing the disintegration of the nation based on the development of technology and information. Therefore, it is necessary to have a reconstruction of civic education become a global civic education that develops students' learning skills and innovation in the 21st century. The concept of 21st century learning is known as the global civic education learning model or global citizens project learning model. According to Reynolds, citizenship knowledge in the citizenship curriculum was usually focused on the national process with less emphasis on the international global structure. Therefore, it is necessary to have a learning model that collaborates between nationalism content and the insight of global citizens (Reynolds, 2012).

These citizen characteristics are used as the basis for the development of a global citizens project learning model that is collaborated on three dimensions of global citizens i.e. global responsibility, global competence, and global participation (Morais & Ogden, 2011). According to Bernadette this concept was used as a media to build "connecting, creating, collaborating, and communicating across a global network," we can scaffold our students' acquisition of the habits of mind and the ways of knowing, thinking, and acting that lead them towards developing global literacy in the 21st century (Castek & Dwyer, 2018).

A global project will build global insight and participation through the use of technology. Technology is the basis for developing learning skills and innovating for global citizens. One form of global civic education that can be done by teachers around the world is through global research that discusses global civic education. Projects that connect teachers around the world can be beneficial not only at a professional and pedagogical level but have proven through research, interactions with people around the world inspire personal reflection because participants learn to do global civic education (Harshman & Augustine, 2013). The concept of global civic education must also have universal values as a form of participation in the realm of global citizens. These values can be sourced from the values of national identity that emphasize the philosophical basis of the character of global society. The characteristics of the global community do not only discuss parts of global problems but also stem from an understanding of global economic and political issues until education policies are increasingly directed towards achieving political and economic goals (Marshall, 2011).

Philosophical basic values of global civic education developed in Indonesia must be sourced from the values of God, humanity, unity, society, social justice, competition, respect for others, independence and peace (Murdiono, 2014) which are then actualized in the Global Citizens Project

Learning Model (GCPL) process on civic education in schools especially at the senior high school level. Through the project learning model, global citizens are expected to be able to develop 21st century student skills, namely learning and innovating skills. One indicator of its achievement can be seen from the perspective of the value of the learning process which consists of social spiritual, knowledge, skills, and literacy competence. This competence is built based on indicators of learning and innovation skills which consist of critical thinking skills and problem solving, communication and collaboration as well as creativity and innovation (Trilling & Fadel, 2010).

The learning process by using the global citizen project learning model can provide experience and basic knowledge for the young generation in solving problems or social conflicts in the community. This is as stated by Johnson that "Activity-driven education that improves social interaction, and thinking, reasoning, and problem solving in the classroom" (Johnson, 2015). One of the main components of the global citizens project learning model is the dimension of citizenship which identifies a conceptual model of global citizenship which includes three dimensions: social responsibility, global competence, and global community involvement (Morais & Ogden, 2011). Social responsibility is the level of perception of interdependence with and social care for others, society, and the environment. One of global competence is to try to understand the norms and cultural expectations of others and utilize this knowledge to interact, communicate, and work effectively outside one's environment actively. Developing global competencies requires recognition of one's own limitations and ability to engage in intercultural encounters. Meanwhile, global community involvement is a demonstration of action or tendency to recognize local, national, and global community problems and respond through actions.

The Global Citizens Project Learning Model is expected to be able to provide insight for young people to take part as global nationals to jointly build a more just and civilized human civilization. The learning model of a global citizens project can be an alternative in the realm of the national education system to jointly build the skills and character of young citizens in addressing the development and use of technology and information wisely so that it will form the character of global citizens who think and have global insight but remain national character and spirit. This is because in the digital era the development of global information technology and global civic education is an important part and needs to be built as a form of global citizen participation, a sustainable future for a world that works together with mutual understanding in world citizenship (Ennals et al. 2009).

The prototype of the global citizens project learning model consists of four components which include (1) determining the topic of the project to be made which includes the impact of climate change, social culture, legal politics, economics and education, (2) creating a global citizens project design that will be made consisting of the objectives, forms and activities as well as material tools, (3) evaluation of global citizen projects consisting of the content of challenges, obstacles, and alternative policies undertaken, (4) publication of global citizens project that can be done on various social media such as Instagram, Youtube, and Facebook (Sutrisno, in prep.).

The learning model of the global citizens project has several characteristics including strengthening the values of Pancasila as the nation foundation, developing learning and innovation skills, and developing the global dimensions of students. The purpose of this learning model is to prepare young people who can participate in solving various issues at the local, national, and international scope and having a global perspective through the use of technology and information.

3.2 *Reconstruction of learning and innovation skills in the global citizens project learning model*

Twenty-first century citizen skills is a major issue in the education system in the current era to prepare young people who have critical thinking and problem solving skills, communication skills, creativity and innovation, and collaboration. Therefore, it is necessary to have the concept of learning strategies that reconstruct learning and innovation skills in the form of learning models (Trilling & Fadel, 2010). The global citizens project learning model is a grand design between learning models that combines the material concepts of global education (global citizenship education) developed by UNESCO with the learning skills and innovating 21st century students. Global civic education developed by UNESCO is the basis of the global citizens project learning model on the basis of building learning skills and innovating citizens of the 21st century. It is reconstructed into five basic components (Eis & Moulin-Doos, 2017).

First, the domain of learning consists of cognitive, socio-emotional, and behavioral. Second, key learning outcomes including activities in which (a) learners gain knowledge and understanding of local, national and global issues and the interrelations and interdependencies of various countries and populations, (b) learners develop skills for critical thinking and analysis, (c) students experience a shared sense of humanity, sharing values and responsibilities, based on human rights, (d) students act effectively and responsibly at the local, national, and global level for a more peaceful and sustainable world, (e) learners develop an attitude of empathy, solidarity, and respect for differences and diversity, and (f) students develop motivation and intelligence to take the necessary actions.

Third, key learning attributes consist of three main activities. They are (a) information and critical literacy which is consists of various activities that include knowing about local, national and global problems, systems and governance structures, understand the interdependence and connections of global and local problems, develop skills for critical inquiry and analysis, (b) relating socially and respecting diversity which consists of various activities which include establishing a relationship of identity and a sense of belonging, develop values and responsibilities based on human rights, develop an attitude to respect and respect difference and diversity, (c) be ethically responsible and involved in carrying out various activities that include appropriate skills, values, beliefs and attitudes, foster personal and social responsibility for a peaceful and sustainable world, and develop motivation and willingness to care for the common good.

Fourth, the topic of global civic education activities consists of nine activities that cover (a) local, national and global systems and structures, (b) issues that affect the interaction and connectivity of the community at the local, national and global level, (c) the underlying assumptions and dynamics of power, (d) different levels of identity, (e) different communities owned by people and how they are connected, (f) differences and respect for diversity, (g) actions that can be taken individually and collectively, (h) ethically responsible behavior, and (i) being involved and taking action. Fifth, education activities explain the dimensions of the age level of global civic education which is taught and can be started from the lower level, namely 5–9 years, the upper level of 9–12 years old, the lower middle level of 12–15 years and the upper secondary level 15–18+ years. Global civic education at the upper secondary level becomes an important part in building the attitudes, knowledge, and skills of young citizens of the 21st century (Eis & Moulin-Doos, 2017).

Therefore, a mechanism for developing global citizens project learning models is needed in developing learning and innovation skills. The form of the learning development path of global civic education through the global citizens project learning model can use the learning development cycle i.e. (a) curriculum planning at this stage is carried out with analysis and planning of materials and learning objectives, (b) learning design includes analysis and planning the place of learning, the purpose of the model being developed, and the learning media used, (c) the development of learning is carried out systematically and structured in order to form the development of effective learning models, (d) learning facilities are used to implement the learning model so that it can be done sustainably, (e) the impact of learning outcomes is obtained by collecting data from the evaluation of learning (Kilfoil, 2008).

The concept of developing a global civic learning model project also originates from the elements of model development. According to Joyce (2009), there were five important elements that described a learning model, including: (a) syntax, which was a sequence of learning commonly called phases, (b) social systems namely the role of students and teachers as well as the required norms, the principle of relaxation that gave the teacher a picture of a procedure, viewed and responded to what students do, (c) support systems namely the conditions or conditions needed for the implementation of a model, such as classroom settings and instructional systems, (d) instructional impacts, and (e) accompaniment impacts. Instructional impact is a learning outcome that is achieved directly by directing students to the expected goals. The accompaniment impact is other learning outcomes produced in the teaching and learning process, as the creation of a learning atmosphere experienced directly by students without direct direction from the teacher (Joyce, 2009).

Thus, the concept of a global citizens project learning model in developing learning and innovation skills must be oriented to the development of the 21st century teacher education curriculum as recommended by Maphosa & Mashau that (a) teacher education curriculum must be redesigned to assist in producing a teacher who can operate in a 21st century environment, (b) the teaching methods used in teacher training must be based on reflection and problem solving to be instilled in the trainees of the same approach teachers should use as teachers after training, (c) a teacher-centered approach that aims at transforming students' knowledge with the concepts of learning centered, acting, and solving problems must be emphasized, (d) information technology must be fully used to improve teaching and learning inside and outside the classroom with or without the teacher's presence (Maphosa & Mashau, 2014).

The global civic education developed by UNESCO has become the basic conceptual framework in developing learning models for global citizens' projects to shape and build learning skills and innovation of students in the 21st century. The Global Citizens Project Learning Model is reconstructed from the results of an in-depth analysis based on empirical facts that occur in the field, especially in students and teachers of Pancasila and Civic Education subjects at senior high school level. Hopefully, the Global Citizens Project Learning Model can contribute to the national education system for building young people who have 21st century skills in the use of information and communication technology based on Pancasila values and can contribute to solving global problems and issues that occur in everyday life.

4 CONCLUSIONS

The learning model of the global citizens project is actualized in the learning process of Pancasila and Civic Education at the levels of senior high school, vocational school, and *madrasah aliyah*. The learning method used is based on student centered learning and oriented towards a scientific approach. The implementation of the global citizens project learning model is implemented in four forms of activities i.e., determining the topic, forming a global citizens project design, evaluation, and publication on social media. The learning model of the global citizens project has the characteristics of strengthening the basic values of the Pancasila as the national foundation, the development of learning skills and innovation, and the development of the global dimension of students i.e., participate to solve various issues at the local, national, and international scope and have a global perspective through the use of technology and information. This study will be the basis for future research to produce a global citizen project learning model for the development of learning and innovation skills.

REFERENCES

Castek, J. & Dwyer, B. 2018. Think globally, act locally: Teaching climate change through digital inquiry. *Reading Teacher* 71(6): 755–761.

Eis, A. & Moulin-Doos, C. 2017. Cosmopolitan citizenship education: Realistic political program or program to disillusion powerlessness? A plea for a critical power perspective within global citizenship education. *Journal of Social Science Education* 16(4): 49–59.

Ennals, R. et al. 2009. Global information technology and global citizenship education. *AI and Society* 23(1): 61–68.

Harshman, J.R. & Augustine, T.A. 2013. Fostering global citizenship education for teachers through online research. *Educational Forum* 77(4): 450–463.

Johnson, K. 2015. Behavioral Education in the 21st century. *Journal of Organizational Behavior Management* 35(1–2): 135–150.

Joyce, B. 2009. *Model-model pengajaran edisi 8.* (A. Fuwaid & A. Mirza, Trans). Yogyakarta: Pustaka Pelajar.

Karsten, S. et al. (1998). Challenges facing the 21st century citizen: views of policy makers. In J.J Cogan & R. Derricott (eds), *Citizenship for the 21st century: An international perspective on education.* London: Kogan Page Limited.

Kilfoil, W. 2008. A model for learning development. *South African Journal of Higher Education* 22(5): 1019–1028.
Kim, B.J. et al. 2011. Civic engagement and internet use in local governance. *Administration & Society,* 20(9): 1–9.
Maphosa, C. & Mashau, S.T. 2014. Examining the ideal 21st century teacher-education curriculum. *International Journal of Educational Sciences* 7(2): 319–327.
Marshall, H. 2011. Instrumentalism, ideals and imaginaries: Theorising the contested space of global citizenship education in schools. *Globalisation, Societies and Education* 9(3–4): 411–426.
Morais, D.B. & Ogden, A.C. 2011. *Initial development and validation of the global citizenship scale.*
Murdiono, M. 2014. Pendidikan kewarganegaraan untuk membangun wawasan global warga negara muda. *Jurnal Cakrawala Pendidikan* 3(3): 349–357.
Reynolds, R. 2012. Civics and citizenship education in its global context: The complexity of global citizenship dialogues. *Education Sciences* 2(4): 190–192.
Sutrisno. 2019. *Data Laporan Hasil Penelitian.* In prep.
Sutrisno. 2019. *Hasil Observasi dan Kuesioner Penelitian.* In prep.
Trilling, B. & Fadel, C. 2010. 21st century skills: Learning for life in our times. *Choice Reviews Online* 47(10): 47-5788-47-5788.
UNESCO. 2015. *Global citizenship education: topics and learning objectives.* United Nations Educational, Scientific, and Cultural Organization.
Toukan, V. E. 2018. Educating citizens of 'the global': Mapping textual constructs of UNESCO's global citizenship education 2012–2015. *Education, Citizenship and Social Justice* 13(1): 51–64.

Empowering Civil Society in the Industrial Revolution 4.0 – Sukowiyono et al (Eds)

Custom amalgamation for civic learning in elementary school: Cultural citizenship literacy viewpoint on Minangkabau ethnic tradition

S.I. Hamid, T. Istianti, M.H. Ismail & F.H. Firmansyah
Universitas Pendidikan Indonesia, Bandung, Indonesia

F. Abdillah*
Universitas Negeri Jakarta, East Jakarta, Indonesia

ABSTRACT: This study aimed to explain teachers' efforts into cultural literacy citizenship of integrating Minangkabau local wisdom to civic education in elementary schools. The research was conducted with a qualitative approach to Minangkabau cultural artifacts such as literature, speech, and customs. The ethno pedagogical approach was also used to examine the substance and cultural strategy of learning by fourth-grade teachers from two schools selected purposely. The research findings illustrated that teachers' success in developing learning with a literacy cultural citizenship orientation was supported by students' cultural homogeneity, teachers as native, and teacher competence. The third finding provides an insight that contextualization, a sense of belonging to a culture, and teacher professionalism need to be considered in developing cultural citizenship literacy-oriented learning. Thus, the study recommends that attention is needed for a multicultural and multiethnic context so that more local and national cultures can be introduced in teacher training institutions.

Keywords: amalgamation, civic learning, elementary school, cultural citizenship

1 INTRODUCTION

The Minangkabau ethnicity's cultural and political identities have been recorded in the history of the Indonesian nation's journey. The nation's founders and national figures from the Minangkabau ethnicity such as Mohammad Mohamad Hatta, M. Yamin, M. Natsir, Sutan Syahrir, Agus Salim, Tan Malaka, to Buya Hamka made significant contributions to the field (Abdullah 2009; De Jong 2012). In education, Mohammad Sjafei through the concept of the Inlander Netherland School (INS) Wood Planting Education in West Sumatra. This background becomes a reinforcing element that cultural literacy-based education development studies can develop with the Minangkabau ethnic background. Social, political, and cultural aspects have positions and implications for the richness of content and learning approaches. In education, the concept of ethno pedagogy, which uses all elements of culture for education, is known as an orientation and foundation for the development of Pancasila and Civic Education (PPKn) learning, Citizenship Cultural Literacy needs to depart from the aspects of the cultural wealth of the Minangkabau ethnicity.

Based on previous studies (Abdillah et al. 2017; Hamid 2016; Hamid & Istianti 2012; Hamid et al. 2019), this research intended to examine idioms and artistic expressions from Sundanese, Javanese, Sasak, Bugis, and Minangkabau ethnicities. Take as an example the case of education in elementary schools where idioms in Sundanese have been projected to be used in an integrated manner in Pancasila and Civic Education subjects (Abdillah, et al. 2017). In order to meet these needs, the enrichment of cultural sources of the Minang ethnic group is taken into account as the

*Corresponding Author

DOI 10.1201/9781003180128-28

prospect of expanding and developing Pancasila and Civic Education learning. Reactualization takes the form of learning models continually being developed and adjusted to actual, measurable, concrete, not merely administrative targets. Moreover, the concept of cultural citizenship is also a dilemmatic partial effort in responding to globalization (Duncker 2015). This is also a form of avoiding negative impacts on social cohesion and government performance (Hlepas 2013). Thus, a Citizenship Cultural Literacy Model was developed with ethnic idioms that were used to enrich learning materials (Abdillah et al. 2017).

As one of the primary and general subjects in schooling institutions, citizenship education has curricular fulfilled its function and role, fulfilling the instrumentation of instilling national moral and political values throughout the history of education and the dynamics of political development itself. Therefore, the emergence of conceptual residues and a reduction in the achievement of meaning in a particular situation and period is possible. However, it can be saved from its latent function as a medium for strengthening and restoring essential values whose sources are stored in the artifacts of local cultural life. The substance of character development departs from public culture (Lee 2013). In the political structures and public policies, it is obligatory to foster community participation and the fostering program is supported by a supportive civic culture for the emergence of citizen involvement (Docherty et al. 2001). Therefore, developing Citizenship literacy based on local traditions is important and strategic in multiplying PPKn learning in schools as political and moral education and modeling and mediation of local cultural products, both homogeneous and integral, based on the diversity of the archipelago.

In this study, we examined the implementation of local wisdom values derived from the customs of the life of the Minangkabau people of West Sumatra. The examination is seen from the perspective of citizenship literacy in the curriculum map in elementary schools and the implementation of the praxis model of citizenship literacy learning based on local Minangkabau traditions in Civic Education and Pancasila at elementary school grade IV–V in Padang City, West Sumatra.

2 METHODS

This study applied the qualitative approach to analyze Minangkabau cultural artifacts. The artifacts include literary works, customs, and idioms. In expressing the analysis results of learning, the ethno pedagogical approach was used to enrich the substance and strategy with a cultural context. The cultural context was then framed with the concept of Moral Citizenship Value (Hamid 2015) as a material for learning. The selection of primary schools was selected purposely to capture the conditions to be understood further. Four teachers were then selected purposely by considering the proximity of the school to the traditional center and being close to urban areas. The data collected was then analyzed using data triangulation to obtain representative and sufficient data to present findings. The teacher was first given training in the implementation of the civic literacy development model, then the teacher immediately practiced it in his PPKn class together with all the learning tools. After this implementation, the teacher was then interviewed. The interview is about the activities he implemented. Then, the students were interviewed on a limited basis to see their impressions of the learning. In the final session of this step, the researcher analyzed the framework of the concept of the moral value of citizenship.

3 RESULTS AND DISCUSSION

3.1 *Implementation of the development of the moral value concept of citizenship based on the cultural typology of the Minang community*

3.1.1 *Description of the content of the NMK concept in Minangkabau culture*

Ethnic communities Minangkabau, as one of those tribes who have a wealth of cultural, the actualization of collective belief and performance behavior everyday both in the homeland and overseas based on the actualization of the social and its role in the dynamics of community life, (economic, political, cultural, and the professions generally) the field livelihoods, can be marked to represent the ethnic group that is strong in preserving the customs and traditions of heritage *ninik mamak* (Abdullah 2009; De Jong 2012). Over time and future development, no doubt the community is experiencing scour changes, a decrease in cohesion above the reference value. The teacher has been viewed as an educator and a member of society that is more educated than the general population. The teacher is a central figure who has a higher literacy level towards the fluctuation of the temperature of civilization and the civilization of society in responding to the challenges to his era (Lee 2013). Teacher against Cultural iteration as the basis of developing life praxis values that can be incorporated into the curricular learning format is strong enough to be believed as the right choice (Docherty et al. 2001).

Table 1 below shows expressions used as the cultural beliefs of the Minang people. These expressions are recommended to being understood in the sense that both the concept and the content-message are raised. After being classified and codified based on the NMK-7 Analysis Model Ichas (Hamid 2015), this classification becomes a content of civic cultural literacy in the design of elementary school Civic Education and Pancasila learning. It is essential to raise them fill out the Development of Citizenship Literacy Teaching Materials into PPKn Learning Designs in Elementary Schools.

Table 1. Recommended local wisdom phrases.

No	Local Phrases	Contents	Classification	Coding
1	*Ombak barayun manuju pantai, riak nyato manuju tapi. Indak guno jadi rang pandai, kalau baulemu indak babudi*	There is no meaning to having intelligence if you are not virtuous	Humble, Know-yourself. Patience, Sincerity: (4)	MK1-1
	Panjang namuah dikarek senteng namuah dibilai, singkek namuah diuleh, kurang namuah ditukuak	Willing to accept advice and acknowledge shortcomings		MK1-2
	Pucuak pauah sadang tajelo, panjuluak buah ligundi, nak jauah silang sangketo, pahaluih baso jo basi	Language ethics and courtesy to avoid cross dispute		MK1-3
	Hiduik batungkek batang bodi, mati bapuntiang ditanah sirah. Jikok pandai bamain budi, dalam aia badan indak basah.	The virtue of having a mind brings advantages to an association.		MK1-4
2	*Nak elok lapangkan hati, nak haluih baso jo basi.*	People always be patient, and people respect acceptable language.	Big-hearted, Honest, Open: (7)	MK2-1
	Nak luruih rantangkan tali, luruih bana dipacik sungguah	Be upright and sincere in relationships, uphold the truth of honesty.		MK2-2
	Nanang saribu aka, haniang ulu bicaro, pikia palito hati, dek saba bana mandatang.	The calmness of thought raises good aspirations, and patience brings truth.		MK2-3
	Nan elok dek awak katuju dek urang, sakik dek awak sakik dek urang.	Do and behave to please the crowd.		MK2-4
	Sabanta sakalang hulu, salapiak sakatiduran	Friendship		MK3-5
	Satali pambali kumayan, sakupang pambali katayo, sakali lancuang kaujian, salamo hiduik urang indak picayo	People will no longer trust one person cheating forever.		MK2-6
	Syarak banamo lazim, adat nan banamo kewi, habih tahun baganti musim, buatan nan usah diubahi.	Obeying a joint decision no matter how difficult and bitter it is to accept it		MK2-7
3	*Nan barek samo dipikua, nan ringan samo dijinjiang.*	always cooperate in doing good	Friendship, loyality: (10)	MK3-1
	Olak olai rang basiang, sorak sorai rang karimbo.	The excitement of working together with joy lightens the work.		MK3-2
	Padanggantiang baranah-ranah, kahilia jalan kapianggu, sasimpang jalan kasikabu Duduak samo randah tagak samo tinggi dalam adat Minangkabau	Does not differentiate human dignity unless the deeds conform to custom		MK3-3
	Partamo banamo Minang, Minangkabau namo kaduo, nan kayo mandi baranang, nan bansaik bandi batimbo	Cooperation and the spirit of sacrifice according to their respective abilities		MK3-4
	Senteang bilai mambilai, panjang karek mangarek	Provide help to friends who are in distress, and provide advice if needed		MK3-5
	Saumpamo aua jo tabiang, umpamo ikan jo aia.	The excellent association helps each other and strengthens and needs each other.		MK3-6
	Tatungkuik samo makan tanah, tatilantang samo minum ambun, tarapuang samo hanyuik, tarandam samo basah.	Good cooperation in society, one heart, and mind		MK3-7
	Titiak buliah ditampuang, maleleh buliah dibaliak.	The result of good cooperation to be enjoyed by many people.		MK3-8
	Duduak sahamparan makan saedangan	Sit in one place, eat the same dish.		MK3-9
	Dimana bumi dipijak disinan langik dijunjuang	Where we are, we can adjust		MK3-10

(Continued)

Table 1. Recommended local wisdom phrases. (*Continued*)

No	Local Phrases	Contents	Classification	Coding
4	*Naiaklah dari janjang, turunlah dari tanggo.*	Always act according to applicable rules, according to custom, religion, and state law.	Respect, obeying the law, being alert, vigilant, (devotion): (7)	MK4-1
	Nak jan jauah panggang dari api, latakkan sasuatu ditampeknyo	Professionalism		MK4-2
	Raso aia kapamatang, raso minyak kakuali, nan bakabek rasan tali, nan babungkuih rasan daun	The spirit of defending family honor		MK4-3
	Suri tagantuang ditanuni, luak taganang kito sauak	customary teachings that are absolutely implemented, without deliberation		MK4-4
	Sandi banamo alua adat, tonggak banamo kasandaran	The wisdom of a traditional house, which itself is a shared truth, the strong support of its customary law		MK4-5
	Siang manjadi tungkek, malam manjadi kalang	We should hold and practice every good lesson and parental advice.		MK4-6
	Tak ujuang pangka mangganai, saragi baliak batimba.	Someone who has complete knowledge and tools, which can be used multi-purpose.		MK4-7
5	*Nan mudo pambimbiang dunia, nan capek kaki ringan tangan, acang-acang dalam nagari.*	Youth, the hope of the nation in the hands of youth lies in the progress and retreat of the nation in the future.	smart, careful, brave, and officer: (3)	MK5-1
	Partamo cupak usali, kaduo cupak buatan. Kalau dulu disasali manjadi tuah pandapatan.	Regret first income, regret then useless.		MK5-2
	Pikia palito hati, tanang hulu bicaro	A thought that has judgment is an antidote. A light illuminates the heart, and serenity will give off good speech.		MK5-3
6	*Ombak ditantang manuju pulau, laia dikambang manantang angin*	To achieve a goal and ideals always experience trials and obstacles	Resilient, steady, hard work, tough (reliable): (6)	MK6-1
	Pandai karano batanyo, tahu karano baguru.	Knowledge is argued for because of learning, education, and asking a lot of questions from people who know.		MK6-2
	Sadang manyalam minum aia, sadang badiang nasi masak	Something work that can be done casually, and does not detract from the work that is being done		MK6-3
	Satitiak jadikan lauik, sakapa jadikan gunuang	Strive with the existing knowledge base to continue to achieve higher knowledge		MK6-4
	Sadang baguru kapalang aja, lai bak bungo kambang tak jadi. Kunun kok dapek dek mandangga, tidak didalam dihalusi.	Every time you demand knowledge, don't break the road, study hard.		MK6-5
	Tabujua lalu tabalintang patah.	To defend the truth, take it with utter persistence.		MK6-6
	Dak ado rotan akapun jadi dak ado kayu janjang dikapiang	Can find solutions at critical times		MK6-7
	Walau kaia yang dibantuak ikan di lauik nan dihadang	We need planning to achieve success.		MK6-8
7	*Ingek di rantiang ka mancucuak, Tahu didahan ka maimpok*	Wise	Fair, wisdom, guarding (Leadership) (5)	MK7-1
	Maukua samo panjang, mambilai samo laweh	Fair		MK7-2
	Nansakik iyolah kato, nan padiah iyolah rundiang. Dek tajam nampak nan luko, dek kato hati taguntiang	The word that hurt was more dangerous than a sharp knife.		MK7-3
	Pado pai suruik nan labiah, samuik tapijak indak mati, alu tataruang patah tigo	Be your ideal young man, calm, decisive, smart, agile, wise, and high-minded.		MK7-4
	Sakalam kalam hari sabuah bintang bacahayo juo.	There are always exceptions.		MK7-5
	Satuntuang tabu dek ulek, satuntuang sajo kito buang	A person makes mistakes, lest all the family be hated.		MK7-6
	Tagak indak tasundak, malenggang indak tapampeh.	A leader who has full authority and authority.		MK7-7
			42	

3.1.2 *Development of citizenship literacy learning message content based on Minang local wisdom in grade IV elementary school*

The development of Citizenship Literacy Learning content refers to the curriculum map for PPKn subjects in elementary schools, formally regulated by the Regulation of the Minister of Education and Culture Number 103 of 2014 concerning Curriculum 2013. Apart from the dynamics that are the place and formal reference for implementing learning in schools, especially concerning PPKn subjects in elementary schools (Abdillah, et al. 2017). As subjects that occupy the foundation and are common to schooling, as in the history of education in the country, as far as change is rationalized as an effort to adjust the strategic achievements of planting moral values within the framework of National Education which is oriented towards the formation of good character of citizens by the philosophy of Pancasila and competence in the sense of being smart and skilled to be able to face the demands of the progress and challenges of the era. Along with the mapping of educational goal achievement points that students must achieve as teaching subjects in schools, the necessary and actual competencies that teachers must have as learning facilitators are an integral part of continuous adjustment and development efforts. In this section, the results of the analysis of the curriculum map in primary schools will be mapped with the demands for necessary skills and innovation of class (IV) teachers in developing conceptual materials and the choice of models and media required.

Table 2. Analysis of the concept of value content and the choice of approach and media.

			Selection	
No.	Material concept	Load moral values	Approach	Media
1	The meaning of the relationship between the Garuda symbol and the Pancasila principles	Aspiring noble, virtuous knights/ good deeds	Integrated thematic	Image/animation Stories, movies Game
2	Implementation of obligations and rights as citizens of society	The habit of being fair	Integrated thematic – scientific	Image/animation Stories, movies Game
3	The diversity of religious communities in society	Tolerance to differences	Integrated thematic	Image/animation Stories, movies Game
4	Forms of ethnic, social, and cultural diversity in Indonesia which are bound by unity and integrity	Togetherness as one nation	Integrated thematic	Image/animation Stories, movies Game

3.2 *Implementation of citizenship literacy PPKn learning based on homogeneous ethnic social culture of Minangkabau, West Sumatra*

In the realm of Minang as the parent of the preservation of the local culture where *Ninik Mamak* is institutionally located, the Minangkabau community can be described as water flows upstream if the Minang people in overseas lands such as on the island of Java show the strength of their customs and language as a reflection of culture. No doubt upstream, especially the city of Padang, West Sumatra, whereas the city of Padang, now as one of the provincial capitals, is like showing progress like other big cities in the country, the inhabitants of Padang city in various professions, including teachers, are no different like the Minang people in Jakarta, Bandung, Yogyakarta, and other cities strongly speak their mother tongue if talk to other people from his village – as *Urang Awak*. However, he also speaks the language of the area where he is, for example, Javanese, Sundanese, Betawi, and standard Indonesian. With this socio-cultural background, there are no obstacles, even arguably in the West Javanese expression "*mobok cadas manggih gorowong*" which means, "finding a way to meet an open door" in mobilizing elementary school class teachers with Minang ethnic backgrounds to apply the Citizenship Literacy learning model based on local traditions in culture and local language, namely Minangkabau.

Even though the Minang realm is known as a hotbed of ethnic migrants (Docherty et al. 2001), with the extensive urban socio-geographical conditions, schools and educational institutions are scattered in the middle and corners of the city. The public and the private sector are racing to provide quality services for the needs of the community. The city of Padang as the center of Government, development of Education and Culture as well as a city of commerce and trade which is open to the entry of other ethnic migrants from all over the archipelago. With these biographical conditions , what teachers do relate to the development of local wisdom-based citizenship literacy is not the only potential for the development of homogeneous values of Minang language and culture (Hlepas 2013). It allows to be integral to the integration stage of diversity values by including a comparison of local wisdom value in the language, songs, and other Indonesian ethnic games (Abdillah et al. 2017; Hamid 2016; Hamid & Istianti 2012; Hamid et al. 2019).

4 CONCLUSIONS

Three prominent factors influence the smooth process of cultural integration in learning in the PPKn subject. The first factor is cultural homogeneity in schools, which is dominated by local people of the same ethnicity, namely the Minangkabau. This homogeneity is supportive because teachers find it helpful to focus on developing teaching and learning materials. The second factor, teachers as indigenous people, so that what is developed is based on authentic cultural experiences in civic learning. The third factor, teacher competence, is the main thing to cultivate and organize the two previous supporters. This research implies that there needs to be a multicultural and multiethnic context so that more local and national cultures can be introduced. Also, teacher training is an aspect that also needs attention.

REFERENCES

Abdullah, T. 2009. *Schools and Politics: The Kaum Muda Movement in West Sumatra (1927–1933)*. Equinox Publishing.

Abdillah, F. et al. 2017. Ethnic Idiom Articulation and Civic Education Material for Elementary School – Development of Cultural Citizenship Literacy. *In 1st International Conference on Educational Sciences –* ICES: 225–229. Bandung: ScitePress.

Duncker, K. 2015. *Cultural citizenship: Understanding the natural states of freedom participation and equality in light of societies self-awareness, -contemplation and -reflection through culture and cultural evolutionary learning by mode of media*. Florence: European University Institute.

De Jong, P.D.J. 2012. *Minangkabau and Negri Sembilan: socio-political structure in Indonesia*. Berlin: Springer Science & Business Media.

Hamid, S.I. 2015. *Nilai moral kewarganegaraan dalam artefak kehidupan sosial kultural masyarakat sunda: Studi eksploratif nilai moral kewarganegaraan dalam ungkapan, artikulasi seni dan ritual adat Budaya Sunda*. Bandung: Sekolah Pascasarjana.

Hamid, S.I. 2015. *Semiotika Pendidikan Kewarganegaraan dalam tradisi lokal Sunda*. Bandung: Rizqi Press.

Hamid, S. I. & Istianti, T. 2018. *Pengembangan pembelajaran PPKn di SD-MI*. Bandung: UPI Kampus Cibiru.

Hamid, S.I. & Istianti, T. 2012. Rekonstruksi nilai moral kewarganegaraan berdasar analisis semantik terhadap ungkapan kultural masyarakat Sunda. *Cakrawala Dini: Jurnal Pendidikan Anak Usia Dini* 3(2): 1–21.

Hamid, S. I. et al. 2019. Mengurai konstelasi filosofis Pancasila melalui literasi budaya kewarganegaraan dan literasi digital kewarganegaraan. *In Prosiding Seminar Pendidikan Nasional Membangun Pendidikan Berbasis Paradigma Higher Order Thinking Skills dalam Konteks Keindonesiaan*. Bandung: UPI Kampus Cibiru.

Hlepas, N. 2013. *Cultural diversity and national performance*. Athens: International Centre for Black Sea Studies (ICBSS).

Lee, T. 2013. *Public culture and cultural citizenship at the Thessaloniki International Film Festival*. Cambridge: Harvard University.

Miller, T. 2002. Cultural citizenship. In E.F. Isin & B. Turner (Eds.), *Handbook of Citizenship Studies*: 231–244. London: Sage Publications Ltd.Permata, A. 2015. Membangun kualitas bangsa dengan budaya literasi. *In Prosiding Seminar Nasional Bulan Bahasa*: 148. Bandung: Universitas Bandung.Hamid, S.I. 2016. A Description of Citizenship Moral Values in Sundanese Society. *Man in India* 96(12), 5125–5141.

Docherty, I. et al. 2001. Civic Culture, Community and Citizen Participation in Contrasting Neighbourhoods. *Urban Studies,* 38(12), 2225–2250.

Empowering Civil Society in the Industrial Revolution 4.0 – Sukowiyono et al (Eds)

Is the 'Tabela' digital book based on Indonesian values as an alternative learning media in the era of industrial revolution 4.0 a solution?

E. Karliani*, Triyani & V.A. Gunawan
Palangka Raya University, Palangka Raya, Indonesia

ABSTRACT: This study aimed to develop digital books based on Indonesian values. This digital book, named "Tabela," could be used as a media solution in learning Civic Education courses. The research procedure starts from the activities of demand analysis, development, testing, product revision, and dissemination. The validation results from the experts showed that the Archipelago Insight material which was integrated with Indonesian values got an average of 4.2 so that the category was good and the average evaluation instrument also reached 4.1 so that the category was also good. In its implementation, the "Tabela" digital book was able to make learning easy, interesting, practical, the material was broad, and could be accessed via gadgets.

Keywords: Dayak culture, digital book, Indonesian value, learning media

1 INTRODUCTION

In the scope of higher education, lecturers as educators are required to be able to change the learning paradigm in the form of patterns and strategies that are more student-oriented through the good use of technology, information and communication (ICT). Referring to the results of problem identification carried out by the research team, several problems were found related to the implementation of civic education courses lectures at Palangka Raya University. (1) Student interest in compulsory courses was generally low. (2) Lecturers who teach civic education courses at Palangka University Raya were not comparable to the number of classes with 54 study programs. The total number of lecturers in the general compulsory subject was 10 lecturers who had to teach two thousand students so the ratio was 1: 200 people. (3) There were no media and learning resources in the Citizenship Education course based on ICT that could be used offline or online by student gadgets.

Referring to those problems, concrete action is needed to be alternative solutions in the implementation of learning, especially in civic education courses at Palangka Raya University.

Taking into account the aspects of student specialization, the potential of lecturers as teaching staff and the availability of other supporting resources, the research team tried to design an alternative ICT-based learning media called "Tabela". In the Dayak language, *Tabela* means youth. In the learning innovative concept of Student-Centered Learning, "Tabela" digital book is a contemporary learning media that suits the souls of young people (Karliani, 2017; Darmaji, et al. 2019; Adi & Muhamad, 2020; Karliani & Triyani, 2020).

"Tabela" digital book is a solution and alternative learning media and learning resource in the era of the industrial revolution 4.0. It applies technological innovations in the learning process and facilitates students to learn independently. It can be accessed via their smartphones independently. It allows students to learn at different places and times, adaptive learning with the equipment they have, according to the abilities and interests of students, there are options that help students learn,

*Corresponding Author

DOI 10.1201/9781003180128-29

provide learning by doing, student centered learning experiences, and better student involvement in the learning process. The use of the "Tabela" application can be collaborated with a blended learning model where lecturers can provide clear learning references, monitor, and add to student learning experiences (Fadel & Trilling, 2009; Danwi, 2016; Darmaji et al. 2019, Dina & Daniar, 2019; Adi & Muhamad, 2020).

The application also contains the noble values of Indonesia as a substance. Indonesian values need to be integrated into the material substance because the Civic Education course has a vision to instill nationalism and love for the country. Indonesian values cannot be separated in the learning process in the classroom so that students have a strong inner bond with Indonesia. The Indonesian values integrated in this research are the values in the Dayak culture (Maarif, 2015; Karliani, 2017; Karliani et al. 2018; Karliani & Triyani, 2020). So, the objective of this research is to develop an application namely "Tabela" digital book based on Indonesian value.

2 METHODS

The study used developmental research (R&D). The stages were market demand analysis, development, testing, product revision, and dissemination. The detailed program of this research activity consisted of (1) developing the substance of the Civic Education Course i.e. Archipelago Insights through a blended learning model that allowed students to have full authority on what they want to learn, of course, adjusted to the achievement of competencies and learning needs of each student, (2) developing problem-based evaluation, problem analysis was obtained based on the assignment to identify factual problems that occurred in society, (3) developing and testing the "Tabela" digital book which was applied to lecture activities in the Civic Education course, (4) the implementation and reflection of the application of the "Tabela" digital book were carried out in small classes for 62 Civic Educations Department students of the third semester who took Civic Education courses.

3 RESULTS AND DISCUSSION

3.1 *Substance development of civic education course*

The 21st century is a century of openness or globalization, where learning content is able to fulfill 21st century skills i.e. (1) learning and innovation skills including mastery of diverse knowledge and skills, learning and innovation, critical thinking and problem solving, communication and collaboration, creativity and innovation, (2) digital literacy skills include information literacy, media literacy and ICT literacy, (3) career and life skills include flexibility and adaptability, initiative, social and cultural interactions, productivity and accountability, leadership and responsibility (Fadel & Trilling, 2009; Danwil, 2016; Wibawa, 2018; Azis, 2018). This research product aimed to accommodate 21st century skills by integrating them with Indonesian values (Dayak culture).

Dayak Unity in the Dayak community is a community to complement each other, ready to serve in the interests of the nation and the country with the philosophy of "Huma Betang," which means we must respect our culture and values. The meaning of *Huma Betang* symbolizes the unifier for living with mutual understanding and upholding human rights. One of the terms that describes the value of Indonesia in Dayak culture is *sapphire kambu tapih kalute kakejau tampayah* (good or bad a person's behavior or words show how little or much knowledge a person has). These values are integrated in the "Tabela" digital book.

The experts of Civic Education and Dayak culture assessed the Civic Education Prototype Substance by focus group discussion. The results showed both in terms of substance and Indonesian values, it got an average rating of 4.2 in the "Good" category. The results related to the development of the substance of the Civic Education material based on input from resource persons and lecturers of the Civic Education course showed that the substance needs to be equipped with case examples depicted through videos, pictures, or descriptions of problems that occur in a society that require problem-solving solutions.

3.2 *The development of learning evaluation instruments with self-evaluation principles*

For the developing learning process, it is not limited to material, methods, and media. However, it needs a measuring instrument that is in accordance with the model being developed based on Higher-Order Thinking Skills (Yulianti, 2018; Rosidin et al. 2019). Evaluation based on factual phenomena in society is shown to stimulate students to think critically and to put more emphasis on students to internalize the values of citizenship. The research team developed learning evaluation with the principle of self-evaluation. They prepared a grid of actual problem cases which would be used as an evaluation instrument.

Higher-Order Thinking Skills-oriented evaluation development aimed to prepare students for having the ability to live as individuals and citizens who are productive, creative, innovative and effective and able to contribute to the life of society, nation, state and world civilization. Educational evaluation experts validate the assessment instrument prototype through focus group discussions. The results showed that the average rating was 4.1 in the "Good" category. So, a problem-based evaluation prototype could be used to measure the achievement of student learning outcomes well.

3.3 *The development of the "Tabela" digital book in the learning process*

The development of information technology can be used in the learning process. Learning that adopts technology is learning that can adapt to the times. The storyboard of the "Tabela" digital book consisted of 16 pages. Cover Page was the initial display of the digital book "Tabela" which contained the title, year of manufacture and the identity of the drafting team. Sub Topics page or homepage contained four sub-topics i.e. The Concept of Archipelago Insights; The Historical, Sociological and Political Sources of Archipelago Insights; Dynamics and Challenges of Archipelago Insights; and the Essence and Urgency of Archipelago Insight.

Page 3 contained a preview of material. Page 4 described the concept and urgency of the Archipelago Insights according to Hasnan Habib, Wan Usman, and the National Defense and People's Consultative Assembly. Page 5 described other meanings of the Archipelago Insights etymologically and terminologically. Page 6 described historical sources of Archipelago Insights. Page 7 described the sociological sources of Archipelago Insights. Page 8 described the political sources of archipelago insights. Page 9 described the concept of Archipelago Insights which originated from the Djuanda Declaration in 1957. Page 10 described the dynamics and challenges of Archipelago Insights. Page 11 described further about the dynamics and challenges of the Archipelago Insights integrated with the Indonesian values of the Dayak people: "*Sapire kambu tapih kalute kakejau tampayah*" which means "good or bad a person's behaviour or words show how little or much knowledge a person has".

Page 12 described the essence and urgency of archipelago insights. Page 13 described the continued sub-topic of the essence and urgency of Archipelago Insights. Page 14 displayed the start page of self-evaluation. Page15 displayed independent evaluation that consisted of 10 varied questions with several alternative answers. Page 16 is a closing page that is filled with the final appearance.

"Tabela" digital book makes the learning process more interesting and easier for lecturers and students alike. Through this application, the ease of Civic Education lecturers is that materials, methods, media, sources and evaluations can be accessed holistically from one application. Second, it accommodated individual learning where students would learn with equipment that was adaptive to their abilities. (Fadel & Trilling, 2009; Danwil, 2016; Wibawa, 2018; Azis, 2018; Dina & Daniar, 2019; Adi, 2020). The digital book "Tabela" facilitated opportunities for distance and independent learning.

3.4 *The implementation and reflection of the "Tabela" digital book in the learning process of civic education courses*

The implementation of "Tabela" digital book in the classroom was carried out on a limited scale in Civic Education Program Study. The class consisted of 62 students through a zoom meeting on

September 20, 2020. The team distributed the link of "Tabela" digital book. Students could open it after successful registration.

After this application was implemented, students reflected the implementation of the "Tabela" digital book in realizing the Indonesian value as a solution alternative learning media in the era of industrial revolution 4.0 through a Google Form. The following table is a reflection of the implementation over "Tabela" digital book.

Table 1. Students' reflection results

No	Response	Expectation	Suggestion
1.	Makes it easy to learn and teach	Can be applied in everyday life	Add more animation and coloring variants that are more attractive and easier to like by today's students
2.	It is very practical and makes learning activities easier	Hopefully it will always be instilled in life, Civic Education learning is very important to learn	It is easier to go to the next page
3.	Incredible	More motivating and encouraging students to learn	Hopefully it will be even better
4.	Good	Become better	The material is further elaborated
5.	Very good and good for the learning process that improves the learning system	Civic Education can be more developed and more advanced in application usage technology	In Civic Education learning using applications, it is best if efforts are made not to use data or internet networks so that all students can access them
6.	Very useful	This application can be implemented in the lecture process	Hopefully it will be improved even more
7.	Very useful, efficient and effective	The hope is that the material published is more reproduced	The e-book is more colorful so it's fun and funny
8.	Excellent	Hopefully it makes it easy for users	Add images related to Civic Education
9.	Very good and can learn more widely	Become greater	The table application is more updated and made as attractive as possible so that those who want to learn it do not feel bored
10.	Interesting because it follows current developments that use technology	Hopefully the technology will continue to advance	Hopefully this application is further improved so that it is easier to learn
11.	Make it easier for students to learn	Hopefully in the future it can be even better and can learn to use this application	The application is further expanded to make learning easier
12.	Very good for students to be fast in doing questions	The hope is that the learning will be easier to understand	Criticism and suggestions on the assessment
13.	Practical and enjoyable	In the future this learning will be of greater interest to all students	Getting better and easier in learning
14.	It is very easy for students to answer and look for material, very useful	increases the speed of responsiveness in learning.	the e-book is good
15.	Very helpful in learning	I can study well and get good grades	it can be used by many people other than students
16.	Provides a fast system that is very complete	It can influence life in the nation and state	Increase it again so that IOS users can also learn to use
17.	Exciting	In the future it will be even better in supporting education	The attached quiz questions are the same but the answers are different
18.	Very helpful, can speed up the teaching and learning process	Increase our motivation for the enthusiasm to learn, expand our knowledge in Civic Education	My suggestion is easier to access, terms of the size is large enough to install

(Resourse: self-processed data)

To achieve 21st century skills, learning trends and best practices must be adjusted. ICT-based learning is no longer just a discourse or a vision, but must be a real action at every level of education, including higher education. Therefore, every lecturer must be literate towards technology and be able to apply it in learning in the classroom (Graham & Dziuban, 2007; Fadel & Trilling, 2009; Danwil, 2016; Wibawa, 2018; Azis, 2018; Dina & Daniar, 2019; Adi & Muhamad, 2020).

4 CONCLUSIONS

"Tabela" digital book based on Indonesian values is an innovation in higher education learning by elaborating on Indonesian media, material and values. The integration of Dayak culture as Indonesian values into material substance of Archipelago Insights material could stimulate the students to become members of a community that complement each other, ready to serve in the interests of the nation and the state. In addition, the students responded positively to "Tabela" digital book positively. It makes learning easy, interesting, practical, extensive material, and could be accessed via gadgets. digital book "Tabela" is an alternative learning solution in the era of the industrial revolution 4.0.

REFERENCES

Adi, R. & Muhamad, S.T. 2020. Technology innovation and learning media in industrial revolution era 4.0 (blended learning) in Physical Education. *International Proceedings Conferences Series*: 120–122.

Fadel, C. & Trilling, B. 2009. *21st century skills: Learning for life in our times*. San Francisco: Jossey-Bass.

Graham, C.R. & Dziuban, C. 2007. *Blended learning environments: Handbook of research on educational communications and technology*. A Project of the Association for Educational Communications and Technology.

Darmaji, D. et al. 2019. *Mobile learning in higher education for the industrial revolution 4.0: Perception and response of physics practicum*. International Association of Online Engineering.

Dina, M. & Daniar, C.A. 2019. The effectiveness of utilizing web learning media towards Islamic Education Learning (PAI) outcome in the era of industrial revolution 4.0. *International Journal of Innovation, Creativity and Change* 8(1): 80–96.

Karliani, E. 2017. *Pengembangan model belajar melayani berbasis nilai sosial spiritual untuk memperkokoh pelibatan kewarganegaraan keindonesiaan*. Doctoral Thesis. Bandung: Universitas Pendidikan Indonesia.

Karliani E. et al. 2018. Huma Betang philosophy as the solidarity prototype and ethnic conflict prevention in Dayak Communities of Central Kalimantan. *Annual Civic Education Conference (ACEC 2018)*: 391–397. Amsterdam: Atlantis Press:

Karliani, E. & Triyani. 2020. Strengthening student's peace-loving attitudes through Huma Betang value-based general education curriculum. *Journal of Human Behavior in the Social Environment*: 1–19.

Rosidin, U. et al. 2019. A combined hots-based assessment/stem learning model to improve secondary students' thinking skills: A development and evaluation study. *Journal for the Education of Gifted Young Scientists* 7(3), 435–448.

Maarif, S.A. 2015. *Islam dalam bingkai keindonesiaan dan kemanusiaan refleksi sejarah*. Bandung: Mizan & Ma'arif Institut.

Wibawa, S. 2018. *Pendidikan dalam era revolusi industri 4.0*. Yogyakarta: UST Yogyakarta.

Yuliati, S.R. & Lestari, I. 2018. Higher-order thinking skills (HOTs) analysis of students in solving HOTs questions in higher education. *Perspektif Ilmu Pendidikan* 32(2): 181–188.

Empowering Civil Society in the Industrial Revolution 4.0 – Sukowiyono et al (Eds)

National *pentigraf* as citizenship education innovation for millennial generations

A.M.A. Shofa*
Universitas Negeri Malang, Malang, Indonesia

A.W. Saputra
Universitas Katolik Widya Mandala Surabaya Kampus Madiun, Madiun, Indonesia

ABSTRACT: Society is in the era of disruption. All fields have been disrupted, including education. Citizenship Education is one of the areas that is susceptible to disruption if it is not packaged innovatively. Internalization of values in Citizenship Education must also be delivered creatively, especially to the younger generation. One way that can be done to innovate citizenship education is through short stories in three paragraphs (*pentigraf*). This study discusses the making of a national *pentigraf* which was initiated by the Pancasila study center State University of Malang (UM) as a pioneer as well as the innovation of citizenship education for the millennial generation. The method of this research is participatory observation research because researchers are involved in providing counseling as well as training in making national *pentigraf*. The results of this study are (1) the national *pentigraf* learning strategy that can be applied by the millennial generation (2) the national collection of *pentigraf* books that have been published and disseminated.

Keywords: *pentigraf*, learning, citizenship

1 INTRODUCTION

The progress of information and communication technology that is increasingly widespread makes today's learners classified as children living in generation Z. According to Tapscot (2008), children called generation Z were born in the era of 1998 to 2009. These children are also referred to as technology kids. They are called generation Z because when they are born, information and communication technology is developing very fast. Television, cellphones, and computers were available when the children were born. Singh (2014) states that generation Z has more enemies than the previous generation. This is because generation Z tends to be more individualistic. Therefore, more understanding is needed to understand this generation born in the 2000 era or generation Z.

The mandate of Law Number 20 of 2003 concerning National Education System Article 37 paragraph 2 states that the higher education curriculum must include citizenship education, the explanation in the article is where citizenship education is intended to form students into human beings who have a sense of nationality and love for the country. But the learning problem of citizenship education in the field as stated by (Rista et al. 2018) is that Citizenship education as a subject of science taught formally to students has increasingly lost "trust" and tends to be underestimated. Moreover, there is an assumption that what is taught in citizenship education is not relevant to everyday life, especially the world of work. Moreover, the assumption is that the material for citizenship education is too complex and philosophical, so it is difficult to understand, especially

*Corresponding Author

DOI 10.1201/9781003180128-30

supported by boring teaching methods. In addition, an understanding of Citizenship Education from an early age (before following it), both concerning the material and teaching methods, is very minimal and even seems insignificant. Students tend to take it for granted especially if students who take it are from technical or exact sciences.

Reality as above is a matter of mutual reflection which is part of a good and bad educational system of learning achievement. educators are not only able to list the demands of learning achievement in students, educators are also required to be able and willing to make corrections to the construction of learning that has been taught. Including adjusting the learning process with the development of the times, the needs and characteristics of students. In addition, learning innovations by utilizing technology are also very important, so learning Citizenship Education is not only theoretical and lectured but also can display videos that contain case studies of daily life so students can be more interested and enthusiastic in participating in learning.

Based on the data exposure, an innovative forum is needed to package citizenship education to be more creative. Besides that, it is also necessary to plant national values in the millennial generation in an interesting and no longer conventional way. One container that can be used is a pentagram or short paragraph three paragraphs. According to Indriadi (2018) *pentigraf* is short for three paragraphs. The conciseness and attractiveness of the stories presented make *pentigraf* as a literary genre that is very suitable to be studied in this era of disruption. Therefore, collaboration between the fields of literature and citizenship is the basis for making this national *pentigraf*.

The purpose of this study is two (1) to describe the national *pentigraf* learning strategy that can be applied by millennial generations and (2) to describe the innovative process of making a national collection of *kebangsan pentigraf* books conducted at the Pancasila Laboratory of the State University of Malang.

1.1 *Pentigraf*

According to Tjahjono (2017), *pentigraf* is actually one of the forms of flash fiction, namely short fiction prose, although there is no standard size related to the brevity. This *pentigraf* was one of the mini-fiction genres that had skyrocketed in the 1980s. One of the pioneers was Tengsoe Tjahjono, a writer who actively sent his *pentigraf* to Suara Indonesia Malang daily (Indriadi, 2018). Besides Tengsoe, of course there are also many national writers who are involved in this field. Unfortunately, the field of *pentigraf* had subsided in the 1990s until the beginning of the 21st century. Other genres began to emerge so as to be able to sink this unique genre. The emergence of *Pentigraf* was only felt lately after several national writers began to descend the mountain to realize the culture of literacy, especially literacy.

In 2015, Tengsoe founded a virtual literary community titled Indonesian *Pentigraf* Village. The community consists of various professional backgrounds. One that unites it is the love of literature and willingness to study literature. The President of the Indonesianigraphy Village is Dr. Tengsoe Tjahjono while the *pentigraf* makers are called *pentigrafis* (Priyanti, 2017). *Pentigrafis* initially linked their three paragraph short stories on Facebook social networks. Each member commented on each other and gave input. The idea to make *pentigraf* as books began to be conceived. The first *pentigraf* book from this community is titled Alea Towards Perfect Robot. The book turned out to be a stimulus as well as a trigger for *pentigrafis* to resonate the culture of writing *pentigraf* in various domains. One of the resonant forms is by making *pentigraf* together with Pancasila Cadres in the Pancasila Laboratory, State University of Malang.

The shape of the *pentigraf* is indeed limited to only three paragraphs, but all the elements of the story must still be presented in a pentagram. These elements are (1) character, (2) background, (3) plot, (4) theme, and other intrinsic elements. According to Budianta (in Tjahjono, 2017), *pentigraf* indirectly invite the reader to explore the deepest essence while extorting one story to just the essence. Therefore, readers will be able to appreciate the essence of the story and this genre is very suitable for even busy readers. Herwanta (2018) said that written works such as *pentigraf* were literary works that could be a means of reporting messages as well as faith. This opinion shows that *pentigraf* can be accepted by all groups.

1.2 *Pentigraf writing techniques*

Short story is one part of literature. Basically, literary works require a creative process in making them, including a *pentigraf.* The length of the story depends on the needs. Sumardjono & Saini (1995) say that the short length of short stories depends on the limitations of the elements. Therefore, the short story is known as intrinsic and extrinsic elements.

In short stories known there are two main elements, namely intrinsic elements and extrinsic elements. Wellek & Warren (1989) say that extrinsic elements are elements outside of literary works that cover history to the environment that forms literary works. Intrinsic elements are the constituent elements of the literature itself, such as characters, plots, settings, and others. One of the most important parts of this intrinsic element is the plot. It could be that the plot is said to be the main frame of the short story. The plot is a plot or storyline. According to Atmowiloto (2013), the plot originated from a story conflict.

The plot is also a determinant of story structure. In general, Kinoyosan (2009) divides three parts of the plot, namely opening, core, and ending. Opening must be written as attractive as possible to attract reading interest. The core is the main conflict in the story. Ending is the end of the story or resolution of the conflict. An interesting ending is an unpredictable ending.

Language in short stories should not be formal or scientific language, nor should language be too flowery so that it cannot be understood by the reader. Shirazy (2008) says that the language in writing good literature is a language that is able to invite dialogue with readers. This has been proven when carrying out the creative process in the form of the Ayat-Ayat Cinta novel. The experience and style of communicative language chosen by Habiburrahman El Shirazy can be used as a reference in writing short stories.

The *Cilukba* strategy is a strategy initiated by today's Indonesian writer, Ayu Utami. The basis of its formation is the theory of pleasure of curiosity (Utami, 2015). Enjoyment of wanting to know is found in the game *Cilukba*. The *Cilukba* strategy has the advantage of being able to make readers curious so they can attract reading interest.

Based on Ayu Utami's narrative in the Short Story Writing Workshop at the Malang City Library in 2014. The applied *Cilukba* strategy was able to improve the ability to write short stories of students in the Utan Kayu community. Some of the prifat writers' training classes which were held individually or in groups also received significant results when successfully implementing the *Cilukba* strategy. This proves that the *Cilukba* strategy has a positive influence in the creative process of writing short stories. Based on these considerations, the *Cilukba* strategy was chosen as a learning strategy to write a *pentigraf* for the millennial generation.

2 METHODS

This type of research is qualitative research. According to C. Edson in Sherman and Webb (1988) argue that qualitative research is a form of moral discourse to understand the past and present. The method used is the method of participatory observation. The participatory observation method is a qualitative research method that involves researchers to plunge while becoming the main actors in data collection and creation. The data used in this study is the national *pentigraf* published by the Pancasila Laboratory of the State University of Malang. Data collection techniques in this study there are three, namely (1) observation, (2) interviews, and (3) literature study. In this study, researchers were also actively involved in informing the publication of *kebangsan pentigraf.* The researcher took part in the training process to make *pentigraf.* There are three stages in making *kebangsan pentigraf,* namely (1) training, (2) making process, and (3) book launching. The three stages were carried out by researchers and young people from the Pancasila Cadre who were fostered by the UM Pancasila Laboratory.

3 RESULTS AND DISCUSSION

This research produced a national *pentigraf* book entitled "I Found the Beauty of Diversity." This book was published in 2019 by the publisher of UM Press in collaboration with the Pancasila Study Center of the State University of Malang. There are 34 *pentigrafs* written by 10 members of the Pancasila Cadre.

The ten *pentigraf* writers are (1) Dona Ashari, (2) Dyan Nurvita Martvianti, (3) Hizam Alif Aditya, (4) Moch Rexy Qolbi, (5) Muhammad Wildan Bahtiar, (6) Muchammad Zulkifli Zuchdi, (7) Siti Saidah, (8) Hafizah Widya Amalia, (9) Izzul Islam Azalina, and (10) Juliana Jasmine. The *pentigraf* titles written by these ten writers are (1) I found the beauty of diversity, (2) Used bicycle for you, (3) Hand phone for Sari, (4) Beautiful tolerance in school, (5) Wrinkled laughter, (6) Obstructed Pole, (7) There are Stories in National Songs, (8) Doubts, (9) Journey, (10) Pipit Twist, (11) Selemah Ranting, (12) Coming soon my blossom bloom, (13) A peacock, (14) Only one friend who knew, (15) A word with a wrong meaning, (16) Humanizing humans, (17) Human crackle goodness, (18) Much acting without much talking, (19) Zainal, (20) Cici, (21) Rescue grass, (22) Mountain and school circles, (23) Unexpected goodness, (24) The beauty of this country, (25) Adit's curious sense, (26) Tirex friendship, (27) Endless struggle, (28) Black superman, (29) Enigma, (30) Great hand, (31) Only so, (32) Right foot, (33) Rawon and satay, (34) Sunday morning blessing.

These *pentigraf* writers come from diverse disciplinary backgrounds so the literacy scheme also varies. This is caused by members of the Pancasila Cadre as the main author, originating from various majors so that the resulting work also varies.

The stories made in the national *pentigraf* are themed about national values. This value is deliberately chosen because it is in accordance with the purpose of making a *pentigraf* which is to instill national values in the millennial generation. The stories told are also fictional, but the authors also do research in their making. The research data is obtained when writers interact in society. The interaction process is what is used in making stories. The stories that were raised were also stories that had been experienced by writers in the community but had not been published. Social facts especially tolerance are the main themes raised in this *pentigraf*.

3.1 *Stage of producing work*

First, the pre-training stage. Before making the book, the Pancasila Cadres were given training first in the Pancasila Laboratory. It aims to provide an initial scheme of *pentigraf* as well as the presentation of techniques for writing short stories of three paragraphs to participants. Before conducting the training, the authors formed a small team to succeed the activity. The small team consists of (1) Ardi Wina Saputra, (2) Abd. Mu'id Aris Shofa, and (3) Chair of the Pancasila Laboratory of the State University of Malang. The first and second names were the researchers themselves, the next was the manager of the Pancasila Cadre and the manager of the Pancasila Laboratory. The initial activities we did were (1) discussing the time and place of the training, (2) formulating the material given to the trainees, and (3) making examples of the national *pentigraf* as the participants' stimulus.

Second, the training phase. This stage is packaged in the form of discussion and writing at one of the cafes at the Pancasila laboratory. Packaging is made interesting so that the creative ideas of the trainees can emerge on their own. There were two sessions of training material, namely the delivery of techniques to write *pentigraf* and the delivery of national content as a form of *pentigraf* writing. After the material was delivered, participants were given two papers. The first paper contains an example of a *pentigraf* and the second paper is blank paper which is used as a tool in writing. Initially, many participants felt difficult and hesitant to try, but the team that had been formed tried to continue to strengthen the participants to be confident in writing. Twenty minutes later, several participants were able to produce one *pentigraf*. Participants who were able to produce a *pentigraf* were asked to read the notes to other participants. This turned out to be able to attract the interest of other participants to immediately complete at least one *pentigraf* in the training. Nearly all participants progressed alternately to read their *pentigraf*. At the end of the training, participants were given two weeks to complete three *pentigraf*. The completed *pentigraf* were collected by email.

Third, the stages of curating, editing, and giving illustrations. *Pentigraf* that have been collected via email are then downloaded by the curator. After that the curator distributes it to the editor and illustrator. The editor does the editing process by communicating to the curator and writer. On the other hand, illustrators have begun working on illustrations in several short stories which are considered necessary to be reinforced with images.

Fourth, the stage of layout and printing. After all the *pentigraf* were edited and some *pentigraf* were illustrated, the next step was the arrangement and printing. This stage is carried out by the publisher. In addition to arranging and printing, the publisher also manages ISBNs and publishes according to the amount desired by the customer.

After the book is published, the steps taken are socialization to the community. The socialization is done by book review. Participants who attended the book review activities came from various circles. The resulting *pentigraf* is perfect for all ages to read and can be taught to all walks of life. *Pentigraf* were also distributed to the school library and used as academic souvenirs as well as alternative readings to instill national values for their readers.

4 CONCLUSIONS

The incorporation of civic and literary education in order to educate the public to increasingly understand the value of nationality, was carried out by the Pancasila Cadre with a team from the Pancasila Laboratory of the State University of Malang. His form is the making of a national *pentigraf*. There are three main stages in making it, namely (1) training, (2) making, (3) publishing. The three stages were carried out by members of the Pancasila Cadre and were curated by the Pancasila Laboratory Team of the State University of Malang. The result of the process is a national *pentigraf* entitled "I Discover the Beauty of Diversity." There are 34 *pentigraf* in this book written by 10 members of the Pancasila Cadre. The technique used in this *pentigraf* is the *Cilukba* technique. The process of dissemination of their *pentigraf* is through print and writing mass media, social networking, and book surgery. It is hoped that *pentigraf* will be able to become a stimulus as well as a trigger for people to want to practice the values of nationality, and to innovate in learning Civic Education.

REFERENCES

Atmowiloto, A. 2013. *Mengarang itu gampang*. Jakarta: Gramedia Pustaka Utama.

Tjahjono, T. 2017. *Pedagang jambu biji dari phnom penh dan cerita-cerita lainnya*. Jakarta: Kosa Kata Kita.

Herwanta, A. et al. 2018. *Surga untuk pohon ulin dan cerita-cerita lainnya*. Sidoarjo: Penerbit Buku Delima.

Indradi, A. 2018. *Kitab pentigraf penjaja kopi tengah hari*. Mojokerto: Temalitera.

Indradi, A. 2018. Pentigraf sebagai penyambung benang putus dalam pembelajaran sastra. *Proceeding Senabasa*. Malang: Politeknik Negeri Malang.

Kinyosan, A. 2009. *Jadi penulis fiksi? Gampang kok!* Malang: Andi Publisher.

Mawarti, R.A. et al. 2018. Us, the society, and our Pancasila: A nowadays service-learning program to prepare the good citizen. *Proceedings of the Annual Civic Education Conference*: 131–134.

Priyanti, B.E. October 4, 2017. Yuk menulis pentigraf! *Koran Surya*.

Sherman, R.R. & Webb, R.B. 2005. *Qualitative research in education: Focus and methods*. London and New York: UK Routledge Falmer.

Singh, A. 2014. Challenges and Issues of generation z. *IOSR Journal of Business and Management* 16(1): 59–63.

Taspcott, D. 2008. *Grown up digital: How the net generation is changing your world*. McGraw-Hill.

Shirazy, H.E. 2008. Tentang menulis karya sastra: Yang dicintai pembaca, yang menggugah minat baca. In Anwar Efendi (Ed.), *Bahasa dan sastra dalam berbagai perspektif*. Yogyakarta: Tiara Wacana.

Sumardjono, J. & Saini K.M. 1995. *Apresiasi kesusastraan*. Jakarta: Gramedia

Utami, A. 2015. *Menulis dan berpikir kreatif*. Jakarta: Kepustakaan Populer Gramedia.

Wellek, R. & Austin, W. 1989. *Teori kesusastraan*. Jakarta: Gramedia.

Empowering Civil Society in the Industrial Revolution 4.0 – Sukowiyono et al (Eds)

Citizenship capability in gender perspective, activists, and specialization of science

M.M. Habibi*, Margono & Sudirman
Universitas Negeri Malang, Malang, Indonesia

M.R.M. Rameli
Universiti Teknologi Malaysia, Johor, Malaysia

ABSTRACT: This study aims to describe differences in citizenship capabilities in the perspective of gender, activists, and scientific specialization of students in East Java. The survey was conducted on 396 students from universities in East Java Indonesia who were drawn proportionally and randomly. Data was collected by questionnaire of citizenship capabilities on a scale of four. The three path ANOVA technique is used to analyze data. The results show that citizenship capabilities differ according to activists, but do not differ according to gender and scientific specialization. To improve citizenship capabilities, citizenship education is recommended so that it more equates students in the activities of campus organizations and the community.

Keywords: citizenship capabilities, gender activist, scientific specialization

1 INTRODUCTION

Citizenship capability is key in efforts to develop good citizens, namely citizens who can carry out obligations, and not just demand their rights. Citizens are required to be able to exercise their positive role in solving problems faced by their communities. Therefore, citizenship capabilities have a broad construct, not just the ability to act, but also include the ability to think critically and deep empathy towards people who are afflicted with problems. Chan's research (2012) shows that citizenship education that combines classroom activities and community activities can provide students with the experience to advocate for social justice.

Research on citizenship capabilities not only illustrates the construct of good citizenship in a democratic setting but also requires an understanding of the factors that contribute to the formation of these capabilities. The construct of citizenship capabilities developed by the Center for Civics Education (CCE) is limited to the ability to make action plans (Corporation for National Service, 2001). The construct is deemed inadequate because it has not provided a real experience in overcoming the problems of citizens. Costa Rica's experience in expanding the scope of citizenship capabilities shows the importance of the ability to cooperate with various parties in solving community problems (Fonseca & Bujanda, 2011). In the Indonesian context, citizenship capabilities are ultimately realized in the ability to work together. The dimension of cooperation came from the agreement as a result of the deliberations they held. Of course, various abilities are needed so that citizens can conduct deliberations. This ability might include scientific and intellectual skills and attitudes toward citizenship.

Citizenship capability is a dependent variable that shows four citizens' abilities, namely the ability to start engaging in public deliberation, the ability to develop quality dialogue, the ability

*Corresponding Author

DOI 10.1201/9781003180128-31

to find thoughtful inquiry, and the ability to work together in solving problems in the community (Fonseca & Bujanda, 2011).

The construction of citizenship capabilities is in line with the opinion of Birdwell, Scott, & Horley (2013). Experience outside of school as a volunteer can improve student capabilities, especially those related to the relationship between work and the social environment. Capability can be defined as the power to do something. Class-based learning successfully notifies students. While service-learning is superior in developing citizenship capabilities, such as teamwork, the ability to hold on to tasks, empathy, keeping time, confidence, and getting along with people of various backgrounds. Service-learning is unique because of its impact on growing social awareness and awareness of citizens. Students feel more able to deal with community problems (efficacy). Increased participation makes students feel aware of the needs of the community, believes that they can make a difference, and is committed to serving the community in the present and later in life.

Urgent things to know are what factors determine the citizenship's capabilities. In the context of traditional Indonesian culture, gender is assumed to play a role in the formation of citizenship capabilities. The interesting thing is that universities living in the realm of modernity should not distinguish between men and women in the capabilities of their citizenship. In the tug of war between modernity and traditionality, gender and political relations in Indonesia become complex which involve cultural, ideological, and historical dimensions (Gadjah Mada University, 2009). This attraction is reflected in Sutarso's (2004) study. He concluded that gender issues in politics led to the rejection of the role of women in politics on the one hand, and support for women's political roles on the other. The problem is how the relationship between gender and citizenship capabilities.

Another factor is community social activities. It is suspected that the high level of community activity is positively related to citizenship capabilities. Research so far has only revolved around the relationship of student activity in school activities (curricular and extracurricular) and citizenship capabilities. Menezes's research (2003) shows that participation experience is generally obtained through student activities in sports and arts organizations, drama, music, or computer clubs. Students also participate in volunteer activities. Activities in schools related to citizenship are Scouts, religious organizations, and the environment. In general, such participation has a positive impact on the concept, attitude, and involvement of citizens. Participating students trust political institutions more, are interested in politics, like social movements in citizenship, and are more positive towards immigrants.

In the Indonesian context, history notes that the student movement reflects their responsibility towards more equitable sociopolitical change (Nugroho, 2013). It can be assumed that sociopolitical activity is positively related to increasing citizenship capabilities.

And the last factor that is thought to be related to citizenship capabilities is scientific specialization. There is no sufficient data that the study program taken by students influences citizenship capabilities. At the beginning of the Indonesian student movement, the activity of medical students in politics was very prominent. The national awakening began with the student political movement. In the colonial era, it was alleged that students with a background in natural sciences and engineering had high citizenship capabilities. Soekarno at that time was an engineering student. However, Hatta is an economics student and Soepomo is a law student. In the era of independence, the relationship between scientific specialization and citizenship capabilities has not been carefully studied.

2 METHODS

This research is classified as a type of survey research because researchers do not treat independent variables. The researcher tried to make a comparison between the citizenship capabilities between men and women; high, medium, and low organizational activities; and specification of scientific and social-humanities sciences.

This research was carried out with the following procedure. First, the researcher designs the construct of citizenship capabilities so that they are arranged sub-variables and indicators in the form of grid tables. Second, the researcher developed a citizenship capability questionnaire based on indicators that had been prepared previously. Third, the questionnaire was tested to analyze its validity and reliability. Fourth, a valid and reliable questionnaire is used to collect data on students of state and private universities in East Java. Fifth, the collected data is analyzed. Sixth, researchers write journal articles and research reports.

The research sample was determined by proportional random techniques and clusters. The sample was 396 students from four regional clusters in East Java, namely Jember, Malang, Surabaya, and Bangkalan. Each region is chosen by two universities (public and private). And each college is represented by two faculties (science and technology and social-humanities). The student population in all the areas studied amounted to 44,753 people. Calculation of sampling using the Solvin technique with a confidence level of 95% (Siregar, 2013).

The instrument of this research is a questionnaire about citizenship capabilities (AKK). The questions in this questionnaire were developed based on the variable construct of citizenship capabilities reflected in the grid. Each question in the AKK questionnaire has four answer choices that show the scale, namely one for the lowest and four for the highest. AKK questionnaire tested its validity with product-moment correlation analysis techniques. The question used in this study has a greater Pearson correlation value with a significance of 0.00. Test reliability using the Cronbach alpha technique. AKK questionnaire was rated reliably with a reliability coefficient of 0.786 greater than 0.6.

The research hypothesis is that the capabilities of the citizenship of East Java students differ according to gender, activists and their scientific backgrounds. Data analysis techniques to test the hypothesis are three-way ANOVA with SPSS version 23.

3 RESULTS AND DISCUSSION

3.1 *Gender perspective*

Of the total respondents as many as 396 as many as 166 respondents were male and as many as 230 respondents were female. The average citizenship capability for male students is 76.11 and women are 77.71. The sig value of Test of Between-Subjects Effects is 0.135. With $\alpha = 0.05$ it turns out $0.135 > 0.05$ so Ho is accepted. The decision is that there is no difference in citizenship capabilities based on gender.

The results of this study can be concluded that gender does not affect citizenship capabilities. This is different from the findings of Bilimoria, et al. (2010). He examines collaborative learning in class management by paying attention to gender aspects, and seeks to use classroom incidents related to gender diversity issues to eliminate discrimination. Gender is one of the problems that must be solved in citizenship education.

The equality of citizenship capabilities between men and women shows that women's representation in political institutions needs to be continuously improved. According to Sutarso's research (2011) the representation of women in people's representative institutions is always fewer, even though female voters are larger than male voters. If their citizenship capabilities are the same, it is necessary to pay attention to social, cultural and religious aspects in increasing women's participation (Arbaiyah, 2011).

3.2 *Activist perspective*

Student activities consist of their involvement in campus organizations, extra campus organizations, community organizations, and political parties. Students take part in campus organizational activities such as the Student Council, Student Representative Body, Student Senate, University Student Executive Board, Department Student Association, Faculty Student Executive Board, Student Activity Unit, and others. The extra-campus organizations attended by students were the Indonesian National Student Association, National Student Front, Indonesian National Student Movement,

Indonesian Christian Student Movement, Islamic Student Association, Islamic Union Student Association (HIMA PERSIS), Indonesian Buddhist Student Association, Muhammadiyah Student Association, FL2MI, Unity Action of Indonesian Muslim Students, Campus Da'wah Institution, Indonesian Islamic Students, Indonesian Islamic Student Movement, Catholic Student Association of the Republic of Indonesia, Indonesian Muslim Student Union (SEMMI), and others. Community organizations include religious organizations, social organizations, hobby organizations, student organizations according to their fields of study, and others.

In the online questionnaire four questions were asked regarding activist variables, each question having the same weight value. Respondents who do not participate in the campus organizations will get a score of one while those who participate will get a value of two. This also applies to other questions related to activists, so the maximum value that students will get is eight, and the minimum value is four. The categorization of student activists is divided into five levels. Respondents who get the accumulated value of four are classified as very low, the respondents who get the accumulated value of five are classified as low, the respondents who get the accumulated value of six are classified as moderate, the respondents who get the accumulated value of seven are classified as high and the respondents who get the accumulated value eight are classified as very high.

Of the 396 respondents, 231 respondents classified as very low activity, as many as 99 respondents classified as low, as many as 28 respondents were moderate, as many as 33 respondents classified as high, and as many as 5 respondents classified as very high. The average citizenship capability for students with very low activity is 75.86, low activity is 77.88, moderate activity is 79.07, high activity is 80.45, and activity is very high is 81.20. The sig value of the Test of Between-Subjects Effects is 0.010. Thus 0.010 is smaller than 0.05, so Ho is rejected. This means that there are differences in citizenship capabilities based on the high and low organizational activities that students follow.

The involvement of students in community organizations influences the improvement of citizenship capabilities. In the context of learning, it seems that citizenship is not enough to be taught in the classroom. Students must be immersed in a broader deliberation arena in the community, so that dialogue in the classroom is expanded with dialogue with community members, the government, and other stakeholders (Cooper, 2008; Fonseca & Bujanda, 2011; Longo, 2013; Nishiyama, 2017). Active citizenship education research in the European Union based on collaborative action is focused on citizenship education carried out by non-governmental organizations (Directorate-General for Education and Culture, European Commission, 2007). Thus, research on mutual collaboration learning is needed which integrates citizenship education in schools and communities as a unit of social change movements.

Capability to participate significantly in social change is not limited to aspects of attitude and knowledge of citizenship. Citizenship capability that is far more important is the power to do things, such as teamwork, the ability to hold on to tasks, empathy, keep time, confidence, and get along with people of various backgrounds (Birdwell, et al. 2013).

According to Nishiyama (2017), children must be given a real experience to become real citizens. Children as actors in democracy are involved in the process of making decisions and solving problems with adults, for example in the form of children's congress forums, community planning, consultative forums, or youth-child parliaments. Children as agents of democracy have an effective role by increasing their contribution to the deliberative system. In other words, the deliberative system provides a space that allows children's unique involvement in democracy, even though their involvement does not significantly affect official government policies.

3.3 *Scientific specialization perspective*

Scientific specialization in this study is the scientific field taken by respondents in public and private universities which are the object of research that are grouped into 2 categories, namely social-humanities and science-technology. Students with science-technology specialties come from the faculties of engineering, Mathematics and Natural Sciences, medicine, and agriculture. And students with social-humanities specialties come from economic faculties, faculties of literature/culture, social/political sciences, and education/teacher science. A total of 202 respondents came from social-humanities specialties. And as many as 194 respondents came from science-technology specialties. The average citizenship capability for social-humanities students is 77.86. And the average citizenship capability of science-technology students is 76.20. The sig value of Test of Between-Subjects Effects is 0.227, far greater than $\alpha = 0.05$. Ho is accepted, so there is no difference in citizenship capabilities based on scientific specialization.

The results of this study can be concluded that scientific specialization does not affect the capability of citizenship. This is in line with Oley's (2012) research which found that although social science faculty students have sufficient knowledge to participate in campus management, they are constrained by campus authorities who do not provide opportunities for their participation.

The results of this study are not in line with the findings of Putra's research (2012) regarding student activists from accounting backgrounds. He concluded that activist students had a social learning style, because of the busy life he had to be able to work with other students to complete his assignments, so he spent a lot of time on campus. Research Putra (2012) does not associate student activists with social and political activities, but provides an overview of social learning styles that are more directed at the ability to work together in citizenship capabilities.

4 CONCLUSIONS

Citizenship capabilities differ according to student activities in East Java, but do not differ according to gender and scientific specialization. Student activities in curricular and extracurricular campus organizations, community organizations, and political parties can improve citizenship capabilities. Thus, citizenship education must involve students in efforts to solve citizen problems through the synergy of campus organization activities and community organizations.

Since citizenship capabilities do not differ according to gender, women's political participation in representative institutions and other public organizations must be proportional. Social, cultural and religious barriers in implementing affirmative policies for women must be overcome.

The scientific specialization pursued by students does not affect the capability of citizenship. Thus, citizenship education which aims to enhance citizenship capabilities can use the same citizenship project model between science and social-humanities students.

REFERENCES

Arbaiyah, P. 2011. *Politik perempuan indonesia. In Margono (Ed.). Pendidikan Pancasila: Topik aktual kenegaraan dan kebangsaan.* Malang: Penerbit UM.

Birdwell, J. et al. 2013 Active citizenship, education and service learning. Education, *Citizenship and Social Justice* 8(2): 185–199.

Bilimoria, D. et al. 2010 Gender in the Management Education Classroom: A Collaborative Learning Journey. *Journal of Management Education* 34(6): 848–873.

Chan, C.K. 2012. Exploring an experiential learning project through Kolb's Learning Theory using a qualitative research method. *European Journal of Engineering Education* 37(4): 405–415.

Corporation for National Service. 2001. Active citizenship" in a facilitators guide to effective citizenship through americorps.

Directorate-General for Education and Culture, European Commission. 2007. Study on active citizenship education.

Fonseca, C. & Bujanda M.E. 2011. Promoting children's capacities for active and deliberative citizenship with digital technologies: The CADE project in Costa Rica. ANNALS, AAPSS, 633: 243–262.

Gadjah Mada University. 2009. *Workshop gender, politik, dan kekuasaan sekolah.* Pascasarjana Universitas Gadjah Mada Yogyakarta.

Longo, N.V. 2013. Deliberative pedagogy in the community: Connecting deliberative dialogue, community engagement, and democratic education. *Journal of Public Deliberation* 9(2).

Menezes, I. 2003. Participation experiences and civic concepts, attitudes and engagement: Implications for citizenship education projects. *European Educational Research Journal* 2(3): 430–445.

Nishiyama, K. 2017. Deliberators, not future citizens: Children in democracy. *Journal of Public Deliberation* 13(1): 56–68.

Nugroho, A. 2013. Peran dan fungsi mahasiswa. Retrieved from: http://catatanaktivismuda.blogspot.co.id/2013/08/peran-fungsi-mahasiswa-pfm.html. (Accessed on June 10, 2020).

Oley, P. C. (2012). *Perilaku politik aktivis mahasiswa fakultas ilmu sosial dan ilmu politik universitas sam ratulangi (suatu studi terhadap pengurus organisasi mahasiswa Fakultas Ilmu Sosial dan Ilmu Politik Universitas Sam Ratulangi tahun 2012).* Manado: Universitas Sam Ratulangi.

Putra, H.G.I. 2012. *Perilaku belajar mahasiswa akuntansi: Aktivis, hedonis dan study oriented.* (Unpublished thesis). Malang: Jurusan Akuntansi Universitas Brawijaya.

Siregar, S. 2013. *Metode penelitian kuantitatif dilengkapii dengan perbandingan perhitungan manual & spss.* Jakarta: Kencana Prenada Media Group.

Sutarso, J. 2011. Konstruksi isu gender dalam politik: studi kasus pemilihan umum 2004. *Komuniti* 2(2): 1–13.

Empowering Civil Society in the Industrial Revolution 4.0 – Sukowiyono et al (Eds)

Development of the *Bahtsul Masail* learning model in the Pancasila and Civic Education subject

A. Mufid*, S. Awaliyah, E. Suhartono
Universitas Negeri Malang, Indonesia

N.A.T. Wulandari
SMP Karya Pagelaran, Malang, Indonesia

M. Mustaricha
Sekolah Menengah Kejuruan Negeri 5 Kota Malang, Malang, Indonesia

ABSTRACT: This study aimed to develop learning models in the subjects of Pancasila and Civic Education centered on learners. Validation and trial results were analyzed using descriptive analysis both qualitatively and quantitatively. Validity test results by material experts obtained a percentage of 93%, model experts by 68%, and learning experts by 88%. The results showed that *Bahtsul Masail's* learning model proved feasible and practical to apply. The response of learners after being the subject of the trial, on average, was 91.3%. That was *Bahtsul Masail* learning model was very attractive learning model. The effectiveness test of the learning model was conducted using *Wilcoxon Signed Rank Test* against *pre-test* and *post-test values*. The test results showed a significance level of 0.000 so it could be concluded that *Bahtsul Masail* learning model was effective to be applied. *Bahtsul Masail's* learning model in the subjects of Pancasila and Civic Education was able to accommodate student-centered learning.

Keywords: learning model, *Bahtsul Masail*, Pancasila and Civic Education

1 INTRODUCTION

Learner-centered learning was proven to improve learning outcomes (Sugiyo et al. 2009, Antika, 2014; Heni et al.2019). However, not all subjects were able to realize such learning, including learning Pancasila and Civic Education (Widiatmaka, 2016; Nurhasanuddin, 2016). Similarly, in Madrasah Aliyah Negeri (MAN) 3 Jombang, the learning of Pancasila and Civic Education (PPKn) has not been centered on students. This could be seen from the tendency of students in grade third science of Madrasah Aliyah Negeri 3 Jombang who were lazy to give reciprocity when learning takes place. Therefore, there needed to be innovation in Pancasila and Civic Education learning (Ikhsan, 2019) by paying attention to the characteristics of learners.

Students of Madrasah Aliyah Negeri 3 Jombang were dominated by *Pondok Pesantren Bahrul Ulum*'s students. In line with the orientation of Pancasila and Civic Education as subjects in school that focused on obtaining the character of the nation as a united, religious, and humanist country (Nurdin, 2015), Islamic Boarding School was very effective educational institution in developing the character of students, especially on humanitarian values that guarantee national unity (Anam et al. 2019). As an educational institution born from local Indonesian wisdom, *pesantren* had typical learning methods such as *halaqah, bandongan, wetonan,* and *sorongan* (Rizal, 2011; Anam, 2017). In addition, there was student's tradition in *pesantren* called *Bahtsul Masail.*

*Corresponding Author

DOI 10.1201/9781003180128-32

Bahtsul Masail was commonly referred to be part of deliberative decision on religious issues (Nasih, 2009). In the process of implementation there was an argumentative debate using systematic logic and proficient rhetoric. The argument would be used as the basis for decision making by *musahhih* (director). The brief showed that in the implementation of *Bahtsul Masail* there was an active role of various parties in solving problems, freedom of speech and refuting each other's opinions, and also participants were required to think critically and argumentatively. In addition, *Bahtsul Masail* also prioritized mutual respect and non-authoritarian behavior (Nasih, 2009).

Based on the study conducted, the activities in the implementation of *Bahtsul Masail* were in line with the characteristics of problem-based learning model which was one form of learning model centered on learners (Chen, et al. 2009; Westwood, 2013). In addition, *Bahtsul Masail* was also a tradition that close to the characteristics of learners as boarding school students. Based on the results of the study, developing *Bahtsul Masail* tradition into a learning model of Pancasila and Civic Education was interesting study to be done. Therefore, this study aimed to develop a model of learning *Bahtsul Masail* was made to be flexibel, interesting, practical, and effective to be applied in Pancasila and Civic Education.

2 METHODS

This research was developmental research using *four-D model modification* developed by Thiagarajan (1974). The research stage included three main things, namely *define, design,* and *develop.* At the definition stage, curriculum analysis, needs analysis, student analysis, model analysis, and learning objectives analysis were carried out. At the design stage began with the preparation of tests, model selection, format selection, and initial design. At the development stage, expert validation and product trials were carried out. The test subjects of *Bahtsul Masail* learning model products were students of third science class of MAN 3 Jombang located in the Pondok Pesantren Bahrul Ulum Tambakberas Jombang.

The test results were analyzed using descriptive analysis both qualitatively and quantitatively. Qualitative descriptive analysis was used to analyze the process of research on the development of *Bahtsul Masail* learning model in the subjects of Pancasila and Civic Education in MAN 3 Jombang, the quality of the learning model obtained from validator inputs, as well as the response of learners during the research implementation process. Descriptive quantitative analysis included data analysis of feasibility, practicality, and effectiveness of learning models using percentage analysis (Arikunto, 2013). Analysing the effectiveness of learning models using *Wilcoxon Test* with the help of SPSS 17.0 software for windows.

3 RESULTS AND DISCUSSION

The products produced in this development research were the *Bahtsul Masail* learning models poured into the typical learning syntax. The learning model was part of a conceptual framework designed systemically to organize a learning pattern, reflecting an educator's methods, techniques, and guidelines in conducting learning activities (Ibrahim, 2017). To measure product feasibility and practicality, *Bahtsul Masail's* learning model was validated by a validator consisting of material expert validator, model expert, and learning expert. Validation results were used as the basis for performing product improvements.

Based on the validation has been done, the material expert concluded the material used in the development of *Bahtsul Masail* learning model was very feasible to use. The assessment of material expert validators covering aspects of content, construction, and language achieved a total score with an average of 93%. Material expert validators recommended expanding the material to suit the scope of problem-solving-based materials. This suggestion was in line with Maryani & Syamsudin's (2009) thinking that learning departed from material scope in the form of contemporary issues, cooperative learning style, and good evaluation, was very effective against the learning outcomes of learners, so that the learning process using the *Bahtsul Masail* learning model was believed to be able to form an interesting learning pattern, especially for students in boarding schools who were studying in formal schools.

Validator model experts assessed that the learning model was feasible and practical to use. This assessment was based on validation results covering aspects of content, construction, and language with an average total score of 68% (Arikunto, 2013). Advice from model experts related to learning steps that must be improved to be more specific. The model implementation steps are in line with three important steps, namely the presentation of the outline of learning materials, group formation, and providing opportunities for learners to explain the material to others (Kurniasih & Sani, 2015).

Validation results from learning expert validators also showed that the learning model was very feasible and very practical to use because the average total score reaches 86% (Arikunto, 2013). Validation from learning experts included assessment of the practicality of the learning model to use, clarity of learning model in supporting the material, suitability of the material in the learning model, suitability of the material to be displayed with the objectives of learning to be achieved, presentation of the manual in accordance with the steps of the learning model, the ability of the learning model in improving student activity, the ease of learning model to understand the material presented, the ease of learning model to increase student knowledge , and flexibility in the learning process.

Since the assessment of the validators showed that *Bahtsul Masail's* learning model was feasible and practical to use, the next step was to test the model in test subject of third science class of MAN 3 Jombang. This was done to determine the level of interest and effectiveness of the learning model. *Bahtsul Masail* learning model in the subjects of Pancasila and Civic Education was carried out through two meetings because the learning model consisted of two activities, namely pre-*Bahtsul Masail* activities and the core activities of the implementation of *Bahtsul Masail*. Pre *Bahtsul Masail* activities consisted of three main activities. First, the division of learners into components *Bahtsul Masail*, namely one moderator, one *muharrir*, one *mushohhih*, and three groups of participants. With this division, *Bahtsul Masail's* learning model accommodated self-learning because the learning experience was organized and controlled by learners (Harsono, 2008). These components had their own duties and some prohibitions. This division of tasks realized good social interaction between learners so that learning became interactive (Harsono, 2008).

Moderators were tasked with leading, maintaining order, arranging and dividing time, discussing the conclusions of the answers agreed by the formulation team, then be offered again to the participants, tapping three times when the problem was considered resolved, and in a state of *dlorurot* moderators can appoint one of the participants to replace it. Moderators were prohibited from participating in opinions, impartial or unobjective, and intimidating participants. The formulation (*muharrir*) was tasked with following the course of the *Bahtsul Masail* event, researching the answers and *ta'bir* submitted, straightening out the answers that were considered deviant, and providing the formulation of answers and *ta'bir-ta'bir* supporters. Formulations were prohibited from imposing answers without any *ta'bir* from participants, speaking before being appointed moderators, speaking out of discussion materials, and disrupting the concentration of participants, such as sleeping, jokes or being emotional. *Mushohhih* was tasked with providing direction and advice to participants and the formulation team and considering and *tashihing bahtsul masail's* decision. *Mushohhih* was prohibited from validating the results of *Bahtsul Masail* before decision was made. Participants were tasked with answering the problem and delivering their *ta'bir* after being given time by the moderator and conveying the text or *ta'bir* to the formulation team. Participants were prohibited from making noises in *bahtsul masail* forums, disagreeing with delegates, and speaking without going through moderators or debates.

Once each component was formed, the educator provided question (*masa'il*) that would be discussed. After that, the students carried out the third activity in pre *Bahtsul Masail*, namely students (groups) conducting research on *masa'il* by looking for answers from references that correspond to the problem. Analysis of the problem might be seen from various factors taking into account the positive and negative impacts (Harist, 2010). This was done with the aim of learners (each component) able to prepare argumentative answers and have clear references. In pre-*Bahtsul*

Masail activities, students were required to actively develop knowledge by conducting research, not only obtaining information from educators (Harsono, 2008).

After the pre-*Bahtsul Masail* activities were completed, at the next meeting was carried out the core activities of *Bahtsul Masail*. The implementation of *Bahtsul Masail* consisted of: (1) opening (*mukaddimah*), (2) visualization of problem exposure by moderators, (3) presentation of answers by participants, (4) categorization/grouping of answers, (5) argumentative debates, (6) enlightenment references by *muharrir*, (7) verification of problems, (8) formulation of answers, (9) ratification by *mushohih*, (10) thanks and apologies from moderators, and (11) evaluation. During this activity, students were required to actively engage both individually and in groups. For example, in argumentative debate activities, each group of participants might work well together (no disagreements). This showed that collaborative and cooperative learning was always invited students to actively discuss (debate) and create in terms of working together to improve critical thinking skills (Harsono, 2008, Yanti, 2014).

After the *Bahtsul Masail* learning model was implemented, the students responded to the learning model's effective use through a questionnaire. Based on the response of learners, it was known that the learning model of *Bahtsul Masail* belongs to very attractive category because the average response of learners is 91.3% (Arikunto, 2013). Related to the effectiveness of *bahtsul masail* learning model, measurements were taken using *Wilcoxon Signed Rank Test*. *Wilcoxon* test was done by testing different results of *pre-test* and *post-test* of learners. The average *pre-test* score of 23 learners was *80.87%* while the *post-test* result was *92.39%*. *Wilcoxon Signed Rank Test* results could be seen in table 1.

Table 1. Result of Wilcoxon signed rank test

	Posttest – Pretest
Z	−4.148b
Asymp. Sig. (2-tailed)	0.000

Wilcoxon Signed Rank Test results in table 1 showed a significance value of 0.000. Since the significance value was less than 0.05, there was a significant difference between *pre-test* and *post-test*. In other words,*Bahtsul Masail* learning model could be used effectively in learning Pancasila and Civic Education. *Bahtsul Masail's* learning model was able to accommodate learner-centered learning activities because each activity reflected active, interactive, independent, collaborative and cooperative learning (Harsono, 2008).

4 CONCLUSIONS

Bahtsul Masail learning model was adapted from the tradition of boarding schools in solving scientific problems. *Bahtsul Masail's* learning model developed in the subjects of Pancasila and Civic Education proved feasible and practical to use. In addition, *Bahtsul Masail's* learning model on Pancasila and Civic Education was attractive and effective to improve learning outcomes because it accommodated student-centered learning activities.

REFERENCES

Anam, S. 2017. Karakteristik dan sistem pendidikan Islam: Mengenal sejarah pesantren, surau dan meunasah di Indonesia. *JALIE: Journal of Applied Linguistics and Islamic Education* 1(1): 145–149.

Anam, S. et al. 2019. The moral education and internalization of humanitarian values in pesantren. *Journal for the Education of Gifted Young Scientists* 7(4): 815–834.

Antika, R.R. 2014. Proses pembelajaran berbasis *student centered learning* (Studi deskriptif di Sekolah Menengah Pertama Islam Baitul 'Izzah, Nganjuk.*BioKultur* 3(1): 251–263.
Arikunto, S. 2013. *Prosedur penelitian suatu pendekatan praktik*. Jakarta: Rineka Cipta.
Chen, et al. 2009. Managing problem-based learning: Challenges and solutions for educational pratice. *Asian Social Science Journal* 11(4): 259–268.
Harist, A.B. 2010. *Islam NU (Pengawal tradisi sunni Indonesia)*. Surabaya: Khalista.
Harsono. 2008. Sudent-centered learning di perguruan tinggi. *Jurnal Pendidikan Kedokteran dan Profesi Kesehatan Indonesia* 3(1): 4–8.
Heni, A.N. et al. 2019. Pengaruh pendekatan pembelajaran *student centered learning* terhadap hasil belajar siswa pada mata pelajaran Al-qur'an Hadits kelas XI di MA Bahrul Ulum Kecamatan Singingi Kabupaten Kuantan Singingi. *JOM FTK UNIKS* 1(1): 143–155.
Ibrahim. 2017. Perpaduan model pembelajaran aktif konvensional (ceramah) dengan *cooperatif* (*make a match*) untuk meningkatkan hasil belajar Pendidikan Kewarganegaraan. *Jurnal Ilmu Pendidikan Sosial, Sains, dan Humaniora* 3(2): 199–210.
Ikhsan, M.A. 2019. Life-based citizenship education textbook for vocational high schools. *Jurnal Ilmiah Pendidikan Pancasila dan Kewarganegaraan* 4(2): 242–249.
Kurniasih, I. & Sani, B. 2015. *Ragam pengembangan model pembelajaran untuk peningkatan guru yang profesional*. Jakarta: Kata Pena.
Maryani, E. & Syamsudin, H. 2009. Pengembangan model pembelajaran IPS untuk meningkatkan kompetensi keterampilan sosial. *Jurnal Metodik Didaktik* 9(1): 1–15.
Nasih, A.M. 2009. Bahtsul masail dan problematikanya di kalangan masyarakat muslim tradisional. *Al-Qanun: Jurnal Pemikiran dan Pembaharuan Hukum Islam* 12(1): 106–129.
Nurdin, E.S. 2015.The policies on Civic Education in developing national character in Indonesia. *International Education Studies* 8(8): 199–209.
Nurhasanuddin. 2016. *Pengembangan bahan ajar Pendidikan Pancasila dan Kewarganegaraan (PPKn) menggunakan model Dick and Carey di kelas X MA Ibnu malik Bangkalan* (Doctoral thesis). Malang: Universitas Negeri Malang.
Rizal, A.S. 2011. Transformasi corak edukasi dalam sistem pendidikan pesantren, dari pola tradisi ke pola modern. *Jurnal Pendidikan Agama Islam - Ta'lim* 9(2): 95–112.
Sugiyo, W. et al. 2009. Efektivitas metode *student centered learning* yang berbasis *fun chemistry* untuk meningkatkan hasil belajar kimia siswa. *Jurnal Inovasi Pendidikan Kimia* 3(2): 469–475.
Thiagarajan, S. 1974. *Instructional development for training teacher of exceptional children*. Minnesota: Indiana University.
Westwood, P. 2013. *Learning and learning difficulties: Approaches to teaching and assessment*. London: David Fulton Publishers.
Widiatmaka, P. 2016. Kendala Pendidikan Kewarganegaraan dalam membangun karakter peserta didik di dalam proses pembelajaran. *Jurnal Civics: Media Kajian Kewarganegaraan* 13(2): 188–198.
Yanti, P. 2014. Pengaruh model pembelajaran kooperatif tipe *team games tournament* (TGT) terhadap kemandirian belajar dan peningkatan kemampuan penalaran dan koneksi matematik peserta didik SMPN 1 Kota Tasikmalaya. *Jurnal Pendidikan dan Keguruan* 1(1): 1–11.

Empowering Civil Society in the Industrial Revolution 4.0 – Sukowiyono et al (Eds)

Pancasila: The Indonesian's source of behavior

E.A. Meinarno*
Universitas Indonesia, Depok, Indonesia

ABSTRACT: This paper departs from the need to implement Pancasila in everyday behavior. The author proposes that Pancasila is included in the realm of psychology, as values, hereinafter referred to as national values. This process was not simple, it began from the idea from Soekarno (1958) to the effort from the first Indonesian psychology scientist, Slamet Iman Santoso (1970). The success of psychology researchers in determining the national values brought impact to studies on its relationship with other psychological concepts. The determination of the national value provided aid for researchers and policymakers to determine which behavior will be built and which are in accordance with it. Further empirical studies are needed, however, most of the founding father's issues have been answered.

Keywords: *Soekarno, Pancasila, values, psychology*

1 INTRODUCTION

Studies of Pancasila as an ideology and politics have been studied and developed since the 1990s until 2000 (Moerdiono, 1991; Poespowardojo, 1991; Alfian, 1991; Somantri, 2006; Ongkokham & Achdian, 2006). Studies of Pancasila as the social capital of society and science has also developed. The effort of scholars to explain Pancasila more scientifically started in the 1980s. Several milestones of Pancasila's scientific studies were by Darmaputera (1982) and followed by Yudi Latif (2011, 2020). Other aspects of Pancasila that have been studied are its value in science development (Setyorini, 2017), Pancasila economics (Mubyarto, 1991), education (Wibowo, 2020), law (Attamini, 1991), social life (Soemardjan, 1991), and culture (Wahid, 1991).

The need to get Pancasila closer as a part of people's life becomes very important. At a macro-social level, the view towards Pancasila, which is considered too far from the community, has been seen. Pancasila, which is perceived as far away, occurs when faced with social inequality that seems to be an unfinished social issue and the behavior of public officials that are not in line with Pancasila (Tobing, 2018). The gap between the idea of Pancasila and reality is slowly being questioned and even a group of people have introduced alternatives to Pancasila. Pancasila began to be opposed to religion, which was initially at the level of discourse until it manifested itself in action. This can be seen in the signs of radicalism, terrorism, and acceptance of transnational ideologies (Arif, 2018; Hartanto, 2017).

This article intends to propose the idea that Pancasila is the source of Indonesian people's behavior. This is because Soekarno, as the extractor of Pancasila, claimed that Pancasila was extracted from Indonesian culture (Soekarno, 1958 [2008]). We can see this excavation claim as an idea that has existed for a long time and has become part of the nation. Thus, Soekarno can be considered to have made a special formula for the formation of behavior that is unique to Indonesia.

*Corresponding author

 DOI 10.1201/9781003180128-33

2 METHODS

This paper was built by exploring the sources of scientific writings that study Pancasila. Another important study material is psychology. As a science that studies behavior and mentality, it is very helpful for translating Pancasila scientifically. Therefore, the argument is built based on literature studies on Pancasila and psychology.

3 RESULTS AND DISCUSSION

Pancasila as a Value Since the beginning of its formulation, Pancasila was placed as the basis of the state and the state ideology. This to some extent affects the difficulty of studying Pancasila so that it can be used daily by the community. Even the New Order regime positioned Pancasila as a myth (Pabottingi, 1977), even though the translation of Pancasila as behavior is increasingly needed, especially since the emergence of movements in the society that tend to destroy the order of society during this decade. One that has emerged is the clash of religious ideas and Pancasila, and the emergence of a state idea in the form of a caliphate that will replace the form of the Unitary State of the Republic of Indonesia.

In Soekarno's speech at the 1945 BPUPKI meeting, he stated that the word "sila" from Pancasila means value. This value is expected to be the basis for the formation of Indonesian's unique behavior. It is unique because Pancasila is extracted from the culture of the Indonesian nation (Soekarno 1958 [2008]). In psychology, the idea of basic behavior is value. The idea of Pancasila as a value was proposed again by Latif (2020). This is interesting because long before that, Indonesian psychology scientists had thought about the same thing.

Slamet Iman Santoso was the first psychological scientist in Indonesia that started the study of Pancasila as a value (Santoso, 1970; Djaelani, 1970). Furthermore, other scientists such as Sartono Kartodirjo called humans with a Pancasila personality as homo pancasilaensis (in Gismar, 2008).

The psychological tradition that tries to study values as part of psychology has been carried out by many scientists (Rokeach, 1973; Scheibe, 1970; Sagiv & Roccas, 2017; Schwartz, 1996). Meinarno and Suwartono have upheld Pancasila as a national value and as a psychological construct (Meinarno & Suwartono, 2010; Meinarno & Suwartono, 2011; Suwartono & Meinarno, 2012). National value is defined as the identification of the acceptance of ideas about what is right, according to expectations and good, along with its opponents (incorrect, not right, not according to expectations and wrong) that is based on Pancasila, which is measured in five dimensions: religious-tolerance, humanity, patriotism-unity, democracy, and social justice. Table 1 presents the definition of each national value (see table 1).

Table 1. The definition of each national value

Pancasila	National Value	Definition
Belief in the one and only God	Religious-tolerance value	Trust in God and carry out His commands according to beliefs and do not impose beliefs on others.
A just and civilized humanity	Humanity value	Recognizing equal rights and obligations, compassion for others, establishing relationships with other nations based on mutual respect.
Unity of Indonesia	Patriotism-unity value	Prioritize the interests of the nation over the self/group, love the homeland and nation, develop a sense of unity for the nation.
Democracy, led by the wisdom of the representatives of the people	Democracy value	Decision making based on deliberation for the common interest by not imposing one's will on others, can be accounted for, and the decisions can be carried out.
Social justice for all Indonesian people	Social justice value	Maintain the balance between social rights-obligations and introspection (in the form of the human's noble quality), and self-development aimed to improve social life.

This result then gave a new identity for Pancasila in the realm of psychology, namely as a national value. Another impact of this effort is the development of the measurement of the five national values, which has rarely been conducted in the development of Pancasila as a scientific study.

3.1 *Value as the basis of behavior*

The enforcement of national values, along with their measurements, helps to explain Pancasila with various other psychological concepts. Several studies include the relationship between national values and citizenship (Meinarno & Mashoedi, 2016), multiculturalism (Meinarno, 2017), patriotism (Putri & Meinarno, 2016) and national identity (Suwartono, et al. 2017).

In other circumstances, national values can also be developed as a basic reference in shaping behavior. The behaviors that are shaped are built with a psychological frame of mind and can be adopted by more parties. This effort can harmonize national values and behavior. The method is to determine which national values will be developed into behavior, then the behavior designer can describe the process and timing of its formation in detail (Meinarno, 2020).

4 CONCLUSIONS

This article responds to the need to realize Pancasila in the form of behavior. This is necessary because of the social situation and conditions that require behavior that supports the progress of the nation. In this case, psychology tries to answer this need by translating Pancasila into national values. This is a breakthrough both scientifically and for practical purposes.

REFERENCES

Arif, S. 2018. *Islam, Pancasila, dan deradikaliasi: Meneguhkan nilai keIndonesiaan.* Jakarta: Elex Media Komputindo.

Attamini, A.H.S. 1991. pancasila cita hukum dalam kehidupan hukum bangsa indonesia. dalam pancasila sebagai ideologi dalam berbagai bidang kehidupan bermasyarakat, berbangsa dan bernegara. (Ed.) Oetojo Oesman dan Alfian. Jakarta: BP-7.

Darmaputera, E. 1982. pancasila and the search for identity and modernity in indonesian society: a cultural and ethical analysis. (Unpublished Dissertation). Boston College-Andover Newton Theological School. Massachusetts.

Djaelani, T. 1970. *Pengetrapan Pantjasila dalam hidup sehari-hari.* Jakarta: Markas Besar Angkatan Darat Sekolah Staf dan Komando.

Latif, Y. 2011. *Negara paripurna.* Bandung: Mizan.

Latif, Y. 2020. *Wawasan Pancasila: Bintang penuntun untuk Pembudayaan edisi komprehensif.* Bandung: Mizan

Gismar, A.M. 2008. Mencari Indonesia. In Hidayat, K. & Widjanarko, P. (Eds.), Reinventing Indonesia: Menemukan kembali masa depan bangsa. Jakarta: Mizan.

Hartanto, D.A. 2017. Implementasi nilai filosofis Pancasila dan agama Islam dalam menangkal paham radikalisme di Indonesia. *FIKRI: Jurnal Kajian Agama, Sosial dan Budaya, 2*(2), 307–344.

Meinarno, E.A. 2017. *Peran identitas etnis, identitas agama, dan identitas nasional yang dimediasi nilai nasional terhadap Bhinneka Tunggal Ika.* (Unpublished Dissertation) Universitas Persada Indonesia YAI.

Meinarno, E.A. 2020. *Watak yang dibangun dari desa untuk Indonesia. Prosiding Webinar Seri 17: Kewargaan (Citizenship).* Kongres Kebudayaan Desa: Pola Relasi Baru Warga dan Negara 2020. Yogyakarta.

Meinarno, E. A & Mashoedi, S. F. 2016. Pembuktian kekuatan hubungan antara nilai-nilai Pancasila Dengan Kewarganegaraan. *Jurnal Ilmiah Pendidikan Pancasila dan Kewarganegaraan* 1(1): 12–22.

Meinarno, E.A. et al. 2019. Isu-isu kebangsaan dalam ranah psikologi Indonesia dalam Psikologi Indonesia. (Eds) Hafiz, S.E & Meinarno E. A. Jakarta: KPIN Rajawali Pers

Meinarno, E.A & Suwartono, C. 2010. The measurement of Pancasila: An effort to make psychological measurement from Pancasila values. *The First International Conference of Indigenous and Cultural Psychology.* Yogyakarta,

Meinarno, EA. Suwartono, C. 2011. The measurement of Pancasila: An effort to make psychological measurement of Pancasila scale. *Jurnal Ilmiah Mind Set.* *2*(2). Juni, 104–110.

Mubyarto. 1991. *Ideologi Pancasila dalam kehidupan ekonomi. Dalam Pancasila sebagai Ideologi dalam berbagai bidang kehidupan bermasyarakat, berbangsa dan bernegara.* (Eds) Oesman, O & Alfian. Jakarta: BP-7.

Ongkokham & Achdian. 2006. Pancasila dari kontrak sosial menjadi ideologi negara. Dalam restorasi Pancasila: mendamaikan politik identitas dan modernitas. (Eds) Nasution, I & Agustinus, R. FISIP UI. Depok.

Pabottingi, M. 1977. *Pancasila dan demitologi [Pancasila and Demitology].* Prisma 8.

Putri, M. A., &Meinarno, E. A. 2016. Studi lintas generasi: Penghayatan Pancasila dan patriotisme pada era globalisasi. *Psikologi Indigenous Indonesia* 2016: Kebhinekaan dan Masa Depan Indonesia: Peran Ilmu Sosial dalam Masyarakat, Malang

Rokeach, M. 1970. *The nature of human values.* New York: The Free Press.

Sagiv, L., & Roccas, S. 2017. *What personal values are and what they are not: Taking a cross-cultural perspective. In Values and behavior.* Springer, Cham.

Santoso, S. I. 1970. Analisa dan interpretasi pantjasila. Jakarta: Markas Besar Angkatan Darat Sekolah Staf dan Komando.

Sastrapratedja, M. 1991. *Pancasila sebagai ideologi dalam kehidupan budaya. Dalam Pancasila sebagai Ideologi dalam berbagai bidang kehidupan bermasyarakat, berbangsa dan bernegara.* (Eds) Oesman, O & Alfian. Jakarta: BP-7.

Scheibe, K. E. 1970. *Beliefs and values. Holt, Rinehart and Winston.* New York: Inc.

Schwartz, S. H. 1996. Value priorities and behavior: Applying a theory of integrated value systems. In The psychology of values. *The Ontario Symposium.* volume 8. lawrence erlbaum.

Setyorini, i. 2018. Urgensi penegasan Pancasila sebagai dasar nilai pengembangan iptek. *Syariati: Jurnal Studi Al-Qur'an dan Hukum* 4(2): 213–222.

Smith, AD. 1991. *National identity.* London: Penguin Books.

Soekarno. 1958. Kursus presiden Soekarno tentang Pancasila di Istana Negara tanggal 16 Juni 1958. Jakarta: PSP UGM dan Yayasan TIFA.

Somantri. GR. 2006. *Pancasila dalam perubahan sosial-politik Indonesia modern. Dalam Restorasi Pancasila: Mendamaikan politik identitas dan modernitas.* (Eds.) Nasution, I & Agustinus, R. FISIP UI. Depok.

Soemardjan, S. 1991. *Dalam Pancasila sebagai Ideologi dalam berbagai bidang kehidupan bermasyarakat, berbangsa dan bernegara.* (Eds.) Oesman, O & Alfian. Jakarta: BP-7.

Suwartono, C., & Meinarno, E. A. 2012. Value orientation scale: The validation of the Pancasila scale. *Jurnal Pengukuran Psikologi dan Pendidikan Indonesia* 1(3): 175–183.

Suwartono, C., et al, G. 2017. Development of the national identity measurement using the implicit association test method. *ANIMA Indonesian Psychological Journal* 32(2): 76–89.

Swasono, S. E. 1991. Demokrasi ekonomi: Keterkaitan usaha partisipatif vs konsentrasi ekonomi. Dalam Pancasila sebagai Ideologi dalam berbagai bidang kehidupan bermasyarakat, berbangsa dan bernegara. (Eds.) Oesman, O & Alfian. Jakarta: BP-7.

Tobing, E. 2018. *Indonesian dream: revitalisasi dan realisasi Pancasila sebagai cita-cita bangsa.* Jakarta: Kompas

Wahid, A. 1991. Pancasila sebagai ideologi dalam kaitannya dengan kehidupan beragama dan berkepercayaan terhadap Tuhan yang Maha Esa. Dalam Pancasila sebagai Ideologi dalam berbagai bidang kehidupan bermasyarakat, berbangsa dan bernegara. Oesman, O & Alfian. Jakarta: BP-7.

Wibowo, B. A. 2020. Pancasila sebagai Landasan Filosofi Pendidikan Karakter Kurikulum 2013. Biormatika: *Jurnal ilmiah fakultas keguruan dan ilmu pendidikan* 6(1): 10–19.

Empowering Civil Society in the Industrial Revolution 4.0 – Sukowiyono et al (Eds)

The effect of perceived parenting styles on civic skills among East Java educators

M. Saputra*, I.H.A. Siddiq, R.P. Prabawangi & S. Untari
Universitas Negeri Malang, Malang, Indonesia

ABSTRACT: The aim of this study was to investigate the correlation between perceived parenting styles and civic skills among Educators in East Java, Indonesia. It was hypothesized that Educators' perception of how they were raised by their parents correlates with their civic skills. We recruited 179 Educators from East Java as our respondents. The respondents answered questions on the Likert scale. Our study found there was a correlation between parenting style and civic skills among Educators, $R = 0.10$, p-value 0,01. In other words, about 90% of civic skills among Educators were influenced by another variable. The authoritative parenting style was a more significant influence on the formation of civic skills than the authoritarian parenting style and permissive parenting style.

Keywords: civic skills, educators, perceived parenting style, parents

1 INTRODUCTION

The rapid development of technology and information has made the countries geographical, political, legal and economic strengths have no clear boundaries, since all of them are connected. This condition is supported by the presence of devices, super-fast internet connection, and cheap internet interconnection that make everyone have no limited space and time. This make it easy for everyone to gain information in online news source, one of them is through social media. (Juditha, 2011; Prisgunanto, 2012; Saputra, 2016; Siswanto, 2018; Watie, 2016).

The influence of the power of the internet and social media on the lives of citizens with the state can be witnessed in the political dynamics of the 2016 presidential election in the United States. The results of research by researchers at New York and Princeton University explain about Americans who pay more attention to education, gender, race, discussion, or the amount more associated with them, which is more proportionate, uses more news stories on Facebook (Newton, 2019). While in England, Jeff McMahan, founder of "The Journal of Controversial Ideas" and professor of moral philosophy at Oxford University in charge of social media is responsible for intolerance in society (King, 2018).

In the Indonesian context, the power of the internet and social media can be witnessed during the DKI Jakarta governor general election in 2017. At that time, a video of alleged blasphemy cases carried out by Basuki Tjahaja Purnama, while conducting a campaign in the community. Due to public pressure on social media, the video was reported by the public and brought him to prison (Al Qurtubi, 2017; Mardira, 2017). Another example during the election of the President and Vice President of Indonesia in 2019, there was a war of discourse in cyberspace in part social media. The term labeling appears in election support groups with animal names. (Stefanie, 2018; Triyoga & Faishal, 2018).

Research finding by Pusat Pengkajian Islam dan Masyarakat Universitas Islam Negeri Syarif Hidayatullah Jakarta (Admin, 2018) with total samples of 2,237 Muslim Educators in 34 provinces in Indonesia explained that 60% of Educators in Indonesia from kindergarten to senior high school

*Corresponding Author

 DOI 10.1201/9781003180128-34

have intolerant and high radicals' opinions. Another study conducted by Indonesian State Intelligence Agency (Akbar, 2018), showed that around 39% of students from three universities were exposed to radicalism. The ease of getting and opening information in the era of the internet and social media is actually a good thing in a democratic country that upholds the principle of free speech as its pillar. It is just that the consequences don't stop there, social media also provides anyone to facilitate negative messages, curses, tricks to facilitate understanding of intolerance and radicalism. So, citizens need a skill to analyze the information that is spread through social media. This skill usually called by civics skill.

A citizens' abilities from identifying until concluding information on politics issues and citizenship is a civic skills definition by Patrick (2002). Another definition of civic skills is the ability of citizens to obtain political information which later be used as a consideration for making political decisions (Maiello, Oser, & Biedermann, 2003). Meanwhile Comber (2005), define civic skills as the ability of every citizen to be able to communicate and think critically about political issues and citizenship. Whereas Kirlin (2005) emphasized that civic skills are the ability of citizens to participate in social life. So, civic skills can be defined as a set of abilities possessed by citizens to understand and participate in political and citizenship issues in their country. This skill includes communication skills, identifying social and political issues and participation in the public.

2 METHODS

Our population is Educators in East Java. Sample in this study was obtain purposively, a total 179 Educators was involved in this study. Data was collected by answering questionnaires about parenting styles developed by the research team consisting of permissive, authoritative and authoritarian parenting. Whereas civic skills, researchers used civic attitude and skill questions (Moely et al. 2002) which consisted of six components namely civic action, interpersonal and problem-solving skills, political awareness, leadership skills, social justice attitude, and diversity attitudes. Furthermore, this questionnaire will be used under the name civic skills. The data was analyzed by SPSS version IBM SPSS Statistics 25. Categorical data was presented in percentage, while numerical data from Likert scale presented in value and average. Correlation between parenting style and civic skills was showed in R square and significant value was defined by <0.05.

3 RESULTS AND DISCUSSION

3.1 *Demographic characteristics*

Among 179 respondents that included in this study, 56.4% was female, 54.2% was range form 23–34 years old and 49.7% was graduates' students. All the data show in table 1.

Table 1. Demographic characteristic of educators

Variables	N (%)
Sex	
Male	78 (43.6%)
Female	101 (56.4%)
Age	
23 – 34 years old	97 (54.2%)
35 – 46 years old	35 (19.6%)
>47 years old	47 (26.3%)
Educational level	
S1	89 (49.7%)
S2	85 (47.5%)
S3	5 (2.8%)

3.2 *Perceived parenting style*

There are 3 perceived parenting style in this study, permissive, authoritative, and authoritarian parenting styles. Our study found that authoritative parenting style have the highest average value, 3.42, followed by authoritarian parenting style 2.86 and permissive parenting style 2.18. Thus, among 179 Educators in East Java define that their parent's parenting style is authoritative.

3.3 *Civic skills*

Civic skills explain how the objective abilities of Educators using self-assessment. There are 30 questions in Likert scale. Average value from 179 respondents is 2.30 (on scale 4). This mean that civic skills among Educators in East Java is moderate categories.

3.4 *Correlation between parenting style and civic skills*

Not only the descriptive results but also correlation between parenting style and civic skills was show in our study. Based on correlation we found that R square is 0.10. This is indicating that parenting style only give 10% of effect in civic skills among Educators in East Java. Other variables give effect as big as 90%. Although only small effect from parenting style the correlation shows statistically significant (p value 0,01). (Table 2)

Table 2. Coefficient correlation

Correlation coefficient	0,32
Critical point	0,19
N	179
Error Level (Two-tailed Test)	0,01
R Square	0,10

4 RESULTS AND DISCUSSION

The practice of citizenship requires an ability that must be possessed by citizens, commonly referred to as civic skills. Civic skill itself is an inseparable part of civic competence. Conceptually, civic competence is wider than civic skills. Civic competence includes the concepts of civic knowledge, civic disposition, and civic skills. In another typology, civic competence consists of the concepts of citizen value, participatory attitudes, social justice, and knowledge and skill democracy (Hoskins, et al. 2015). Gerosa (2019) did not give a clear typology contained in civic competence. But his presentation of civic competence was the same as another conception. Civic competency in general is a competency that must be possessed by citizens, which contains values, knowledge, attitudes and skills. Through this civic competence, citizens are directed to have knowledge about the political and legal system in the country where they live. they must also have values that they believe are in accordance with the basic ideology of their country, to the point of attitudes and behavior that are in accordance with values that have been mutually agreed upon in society.

Once the importance of this Civic Competence variable presents several questions such as, what are the factors driving the increase or decrease in civic competence from various civic competence models and typologies, which are the highest concerns? For the second question, there is no research that explains which is more important than the various concepts contained in civic competence. However, it seems that civic skills are interesting enough to be presented. For this reason, the next step is how the second question can be answered. What are the factors that influence the existence of civic skills? How does it affect This study provides an explanation that the one that affects civic skills is the concept of perceived parenting style.

This research focuses on the causes of the formation of civic skills among Educators in East Java. Various things can be the cause of the formation of the civic skills. But on the other hand civic skills also have an impact or effect on other phenomena such as the existence of good civic skills among citizens has an influence on political participation (Jeong, 2013). Civic skills can also be influenced by other phenomena. For example, such as the integrity of the relationship between the parents affect a child's civic skills. Children with divorced parents have lower civic skill abilities than children whose parents are still together (Voorpostel & Coffé, 2015). This is certainly related to the socialization obtained by children from their parents.

The results of this study support the study, but in the context of parenting style. However both of them pay attention to the presence of parents who have influence on the formation of civic skills in the future. This study found that the existence of civic skills is influenced by parenting style. When in one's childhood he is raised in an authoritative style so in the future parents will see their children grow up with sufficient civic skills.

5 CONCLUSIONS

Based on research findings that have been presented previously, it is found that perceived parenting style is positively and significantly related to the weak category of civic skills among Educators in East Java. It was also explained that the authoritative parenting style was the most significant influence on the formation of civic skills compared to the authoritarian parenting style and permissive parenting style. With regard to the results of this study, research on weak parenting is due to parenting obtained by Educators that have occurred in the past. As we get older, more and more factors can be used to change parenting towards civic skills.

REFERENCES

Admin. 2018. Survei PPIM 2018: Menyibak Intoleransi dan Radikalisme Guru. Retreived from: https://conveyindonesia.com/survei-ppim-2018-menyibak-intoleransi-dan-radikalisme-guru/

Akbar, W. 2018. BIN Ungkap 39 Persen Mahasiswa Terpapar Radikalisme. Retreived from: https://www.cnnindonesia.com/nasional/20180429023027-20-294442/bin-ungkap-39-persen-mahasiswa-terpapar-radikalisme

Al-Qurtubi, S. 2017. Manipulasi Agama Dalam Pilkada Jakarta. Retreived from: https://www. dw.com/id/manipulasi-agama-dalam-pilkada-jakarta/a-38379477.

Gerosa, T. 2019. Measuring adolescents' affective civic competence: Validity and cross-group equivalence of a second-order hierarchical latent construct. *Applied Research in Quality of Life* 14(2): 335–358.

Hoskins, B. et al. 2015. Civic competence of youth in Europe: Measuring cross national variation through the creation of a composite indicator. *Social Indicators Research* 123(2): 431–457.

Jeong, H.O. 2013. From civic participation to political participation. *VOLUNTAS: International Journal of Voluntary and Nonprofit Organizations* 24(4): 1138–1158.

Kirlin, M. 2005. Understanding the relationship between civic skills and civic participation: Educating future public managers. *Journal of Public Affairs Education* 11(4): 305–314.

King, S. 2018. Social media is responsible for 'increasing intolerance' of ideas, says Oxford professor. Retrieved from: https://talkradio.co.uk/news/social-media-responsible-increasing-intolerance-ideas-says-oxford-professor-18111328762

Maiello, C. et al. 2003. Civic knowledge, civic skills and civic engagement. *European Educational Research Journal* 2(3): 384–395.

Moely, B.E. et al. 2002. Psychometric properties and correlates of the civic attitudes and skills questionnaire (CASQ): A measure of students' attitudes related to service-learning. *Michigan Journal of Community Service Learning*, 12.

Newton, C. 2019. *People older than 65 share the most fake news, a new study finds: And the finding holds true across party lines.* Retrieved from: https://www.theverge.com/2019/1/9/18174631/old-people-fake-news-facebook-share-nyu-princeton

Oskooii, K.A.R. & Dana, K. 2018. Muslims in Great Britain: The impact of mosque attendance on political behaviour and civic engagement. *Journal of Ethnic and Migration Studies* 44(9): 1479–1505.

Owusu-Agyeman, Y. & Fourie-Malherbe, M. 2019. Students as partners in the promotion of civic engagement in higher education. *Studies in Higher Education*: 1–15.
Patrick, J. J. 2002. *Defining, delivering, and defending a common education for citizenship in a democracy*. 36.
Saputra, M. 2016. Pengaruh habituasi, media sosial dan pembelajaran pendidikan kewarganegaraan terhadap kesadaran lingkungan siswa sma negeri se-kota Bandung. *Jurnal Ilmiah Mimbar Demokrasi* 15(2):23-27.
Siswanto, T. 2018. Optimalisasi Sosial Media Sebagai Media Pemasaran Usaha Kecil Menengah. *Liquidity*, 2(1): 80–86.
Watie, E.D.S. 2016. Komunikasi dan media sosial (communications and social media). *Jurnal The Messenger* 3(2): 69.

Empowering Civil Society in the Industrial Revolution 4.0 – Sukowiyono et al (Eds)
© 2021 copyright the Author(s), ISBN 978-1-032-01801-0.
Open Access: www.taylorfrancis.com, CC BY-NC-ND 4.0 license

Author index

For Product Safety Concerns and Information please contact our EU representative GPSR@taylorandfrancis.com
Taylor & Francis Verlag GmbH, Kaufingerstraße 24, 80331 München, Germany

www.ingramcontent.com/pod-product-compliance
Ingram Content Group UK Ltd.
Pitfield, Milton Keynes, MK11 3LW, UK
UKHW052212190926
475702UK00006B/976
* 9 7 8 1 0 3 2 0 1 8 0 1 0 *